AF251228

Distributed Morphology Today

Distributed Morphology Today

Morphemes for Morris Halle

Edited by Ora Matushansky and Alec Marantz

The MIT Press
Cambridge, Massachusetts
London, England

MIT Press books may be purchased at special quantity discounts for business or sales promotional use. For information, please email special_sales@mitpress.mit.edu or write to Special Sales Department, The MIT Press, 55 Hayward Street, Cambridge, MA 02142.

This book was set in Syntax Times Roman by Toppan Best-set Premedia Limited. Printed and bound in the United States of America.

Library of Congress Cataloging-in-Publication Data

Distributed morphology today: morphemes for Morris Halle / Edited by Ora Matushansky and Alec Marantz.
   pages ; cm
"This Festschrift honors Morris Halle, on the occasion of his 90th birthday, for his many foundational and lasting contributions to our understanding of morphology."
Includes bibliographical references and index.
ISBN 978-0-262-01967-5 (hardcover : alk. paper)
 1. Grammar, Comparative and general—Morphology. 2. Linguistics. 3. Philology. I. Matushansky, Ora, editor of compilation. II. Marantz, Alec, editor of compilation. III. Halle, Morris, honoree.
P241.M596   2013
415′.9—dc23
                                                                    2013000605

10 9 8 7 6 5 4 3 2 1

Contents

Morris, Distributed: An Introduction

Alec Marantz and Ora Matushansky

This Festschrift honors Morris Halle, on the occasion of his ninetieth birthday, for his many foundational and lasting contributions to our understanding of morphology. Morris projects such a giant presence in linguistics in general and in morphology specifically that we, as his students, felt we could only attempt to honor a small, particular chapter in his linguistic life, that of a founder of the theory of Distributed Morphology. From his earliest work on Russian and English phonology, Morris has made major advances in morphology and has advised PhD students on work that has proved foundational in the field—in particular Mark Aronoff (1974) and Shelly Lieber (1980), who continue to stand at the forefront of morphological thinking (see Lieber 2010 and Aronoff and Fudeman 2011). But through the last two decades, Morris cultivated a new flowering of research in morphology. The contributions to this book are written with the deepest appreciation for his role in leading us, his students, to new discoveries and now to a rich tradition within Distributed Morphology.

While many of the particulars of the theory of Distributed Morphology grew from discussions between Morris and one of us (Marantz) starting around 1990, the seeds of the theory were of course planted in Morris's thinking, as summarized in, for example, the *Prolegomena* (1973), from his International Congress of Linguists address. In Marantz's contribution to this book (chapter 6), he identifies as crucial to the birth of DM a dispute over the role in the grammar of what Morris called "abstract" morphemes, or "Q"s. What emerged from these discussions was a radically antilexicalist theory that joined two core assumptions: (1) syntactic approaches to word structure most importantly championed by Shelly Lieber and (2) a rejection of a phonologized set of morphological pieces stemming from "realizational" theories of morphology like those of Robert Beard (1995) and Steve Anderson (1992). Animated discussions with students have always been central to Morris's research, and the early development of Distributed Morphology was also shaped by the work of Eulàlia Bonet (1991) and Rolf Noyer (1992), who were writing dissertations at the time that Morris and Alec were arguing about Qs. Morris encourages his students to confront large, intricate sets of data in all their glorious complexity, to appreciate the devil in the details, and to attempt a full analysis, leaving no generalization unaccounted for and no exception unremarked. Eulàlia had as her task the morphophonology of Romance clitics, with a complete analysis of Catalan, while Rolf was examining Semitic agreement morphology, from both a synchronic and a diachronic perspective. The Distributed Morphology approach to the major issues in morphological theory—syncretism, allomorphy, and blocking—was developed by Morris, Alec, and their students as theory was confronted with new sets of data, including those in Eulàlia's and Rolf's dissertations.

A key to Morris's success and that of his students has been his insistence on the completeness of any analysis of a phenomenon, with a circumscription of the relevant data that might strike some as involving an (over)abundance of technical detail. Besides benefiting from Eulàlia's and Rolf's dissertation research, the theory of Distributed Morphology was built on analyses of Potawatomi and Georgian inflectional morphology, where the spur to account for these paradigms came from Marantz and Halle's need to explain how a "morphous" (morpheme-based) approach to morphology improves the accounts of these data within a realization theory, arguing against Steve Anderson's A-Morphous analysis of the same facts. Morris insisted on a thoroughness in the generation of the Potawatomi paradigms well beyond what was necessary to make the theoretical point. But it was only this discipline of completeness that offered new insights into what was going on in that language and allowed Distributed Morphology to advance Algonquin studies more generally.

Morris's general style of interaction with his students most certainly also contributes to his and their success in advancing our knowledge of morphology. Some of us start working with our students by accepting as true their analyses of the important details they bring to the office; Morris often insists outright that a student's idea has to be wrong. It's up to the student, with Morris's prodding, to refine and improve his or her analysis. As all of us could testify,[1] Morris's enthusiasm and generosity in his meetings with students leave them feeling inspired and important; we've all gone eagerly back to our drawing boards from these encounters. Eventually, Morris is telling everyone about the brilliant discovery his student has made.

This book, then, might be titled "Distributed Morphology in Progress," as much as "D.M. Today," with the hope that, once more, Morris will find our contributions interesting enough to inspire new generations of morphologists.

From the first set of students in the Distributed Morphology era on, Morris's support, input, advice, and exuberant enthusiasm have encouraged research, exemplified in this book, that is converging on a theory of word structure completely embedded within a general theory of grammar. With Morris's spirit, and in his honor, these contributions strikingly demonstrate the Hallean principle that a theory of Morphology is a theory of Everything, and we are delighted to present them as a tribute to our teacher, mentor, and friend.

Readers will note a strong coherence among the chapters, with extensive overlap of questions, assumptions, and approaches that goes beyond the mere sharing of a general theoretical framework. Whether it is the nature of phases, the notion of a morphosyntactic feature, allomorphy and exponence, the synthetic/analytic alternation, stress assignment, or syntactic agreement that is explored, each chapter relies on and highlights the tight connection between morphology and other grammatical modules at the core of Distributed Morphology. It is therefore not surprising that most contributions to the book address or touch on issues in locality, particularly locality at the phonological and semantic interfaces as potentially determined by the syntactic structure of words. Having ordered the chapters counteralphabetically to reverse the usual bias, here we would like to emphasize particular shared topics, though other arrangements might bring out a different set of resonances.

Bringing the foundational issues to the forefront, the relevance of the underlying structure for the interface between morphology and phonology is highlighted in the contributions by David Embick, Tatjana Marvin, and Rolf Noyer.

Thus David Embick demonstrates in chapter 9 that nonaffixal morphophonological alternations (e.g., *sing/sang*), which are often thought to support affixless approaches to morphology (e.g., Anderson 1992, Aronoff 1976, or Stump 2001), show locality effects that are expected in a theory where they are analyzed as linked to (null) affixes: their effect is systematically morphologically and phonologically localized to a hierarchical and linear position within the complex word. Thus, for instance, irregular stem allomorphy in the Italian *passato remoto* only occurs in the absence of the theme vowel-that is, when the (null) past-tense morpheme triggering the stem change is linearly adjacent to the stem (Calabrese 2012). Further evidence is provided from German umlaut, Terena first-person singular nasalization, Chaha masculine-object labialization, and Ischian second-person singular metaphony. Affixless approaches to morphology are shown to have problems with this local character of morphophonological alternations, since they do not postulate a hierarchical structure that would provide a locus for the trigger of morphophonological change.

The role of morphology for phonological analysis is elegantly demonstrated for Proto-Indo-European stress assignment by Rolf Noyer in chapter 2. By combining the hypothesis that individual morphemes may be specified for accent with the options made available by the Simplified Bracketed Grid Theory (Idsardi 1992), Noyer derives the five traditional accentuation classes of Proto-Indo-European stems. In particular, the most accentually problematic stem type, the proterokinetic, is derived through a combination of accent pre-specification and the novel property of certain morphemes to render unaccentable any following syllable; the remaining four classes are derived by more standard means.

Tatjana Marvin also examines the relevance of word structure for stress assignment in chapter 5, which reconsiders the thorny topic of English stress. Marvin argues that in order to account for the preservation of stress and vowel quality in English affixation, stress assignment mechanisms must make reference to the internal structure of derived words, contra surface-based OT accounts. She demonstrates that the behavior of "mixed suffixes" and multiple suffixation, attributed by Burzio 1994 to a metrical consistency hierarchy and coincidence, respectively, is naturally explained in a phase-based approach.

Expanding on the mere fact of the relevance of syntactic hierarchy for morphophonological processes, the issues of cross-modular locality in morphology are examined in the chapters by Jonathan David Bobaljik and Susi Wurmbrand, Heidi Harley and Mercedes Tubino Blanco, and Alec Marantz, which all connect to the syntactic notion of a phase.

In chapter 11, Jonathan David Bobaljik and Susi Wurmbrand offer an ambitious link between the syntactic and the morphological notions of locality: the syntactic phase and the morphological cycle. Building on the assumption that a cyclic head triggers the spell-out of its sister, Bobaljik and Wurmbrand argue that the formation of a cyclic domain is suspended if the cyclic head Y depends for its interpretation on the head X taking YP as its complement. Evidence for this generalization comes from suppletion in superlatives (possible only if the corresponding comparative is also suppletive) and from QR out of embedded clauses (only possible when the tense or the mood of the clause—i.e., the value of its highest head—is determined by the embedding head).

The nature of allomorphy is analyzed by Heidi Harley and Mercedes Tubino Blanco in chapter 7, dedicated to the arbitrary morphological classes in Hiaki (Yaqui). Examining the distribution of bound and free allomorphs of Hiaki lexical stems, Harley and Tubino Blanco argue that class features are not properties of roots, but rather of Vocabulary Items (i.e., of the phonological exponents inserted at the end of the syntactic derivation). Importantly, bound forms differ in this respect from true suppletion, where the two allomorphs are not phonologically related and may belong to different morphological classes. It is furthermore shown that readjustment rules deriving bound forms from free forms are cyclically conditioned: in the verbal domain they apply to all stems and affixes closer to the root than passive and future/irrealis markers and therefore can be argued to belong to the vP (voiceP) phase.

While locality conditions on allomorphy have been the subject of many investigations, conditions on allosemy (i.e., the choice of one of the set of meanings of a particular root) have hardly been studied. In chapter 6, Alec Marantz argues that allosemy is constrained by the cycle in exactly the same way allomorphy is. Support for this view comes from the facts usually cited as problematic for the equation of a "phase" with the domain of special meanings: Japanese nominalizations (Volpe 2005), Greek stative participles (Anagnostopoulou and Samiotti forthcoming), and English stative participles, which are all argued to involve a semantically null v, in full parallel to the contextual allomorphy of the root over a phonologically null v in the English past tense.

Issues of locality also arise in the chapters focusing on the realization of functional morphemes. Exponence is studied by Martha McGinnis and by Karlos Arregi and Andrew Nevins, while the binary versus privative nature of features, the building blocks of such functional morphemes, is investigated by Daniel Harbour.

In chapter 3, Martha McGinnis examines multiple exponence and fission in Georgian number marking to argue in favor of a Distributed Morphology account over Anderson's proposal couched in the terms of A-Morphous Morphology. McGinnis attributes to syntactic competition the impossibility of having more than one number-marking suffix: the subject and a [participant] object compete for a plural-number feature on T (Béjar 2003), ensuring that only the highest plural argument triggers agreement. The ostensible exception, the dative first person plural agreement, is accounted for by assuming that in Georgian, as in a number of other languages, the apparent dative first-person plural is actually a separate person, a collective singular first person, and as such does not compete for plural-number agreement.

Vocabulary Insertion under the conditions of underspecification is examined by Karlos Arregi and Andrew Nevins in chapter 12, once again linking morphology to phonology in their detailed investigation of the Elsewhere Principle. Drawing on Basque pronominal clitics and Bulgarian definite articles, they argue that more underspecified Vocabulary Items may take precedence over more specific lexical entries on the condition that the former have a richer contextual specification. Thus in Basque, case-neutral proclitics can only be inserted in certain contexts and therefore in these contexts take precedence over case-marked enclitics, which have a less specific contextual restriction despite having a greater number of features matching the terminal node. Likewise, the realization of the Bulgarian definite clitic, argued to be sensitive to both phonological and morphosyntactic factors, is shown to be determined by the former (specifying the context for insertion) in preference over the latter (providing the featural specification only).

A core concept shared between all modules of grammar is that of features. In chapter 8, Daniel Harbour sheds new light on the nature of morphosyntactic features, arguing that privative features ([F]) are not sufficient and bivalent features ([+F], [−F]) must be adopted. Diverse crosslinguistic evidence is provided in favor of this conclusion, ranging from gender in Kiowa-Tanoan to number in Bininj-Gunwok and person in Tibeto-Burman. One class of arguments comes from the analysis of number as the recursive composition of the bivalent features [±minimal] and [±bounded]—a treatment impossible with privative features. As a result, Harbour can derive complex number specifications (unit augmented, trial, great paucal) as straightforwardly as the more familiar singular, dual, and plural; for instance, trial is derived as +minimal (−minimal (−atomic (noun))). Binary features also account for the composed plural of Damana and analogous patterns within person systems, such as the composed exclusive of Limbu. Further arguments are provided by alpha exponents (realizing two covariant features at once) and a novel treatment of the morphosyntax of objects in the Kiowa-Tanoan language Tewa: by claiming that animate third-person NPs are [-participant] (as opposed to inanimate NPs, which are unspecified for [±participant]), the chapter derives a variety of phenomena previously unnoted, or unexplained, within Tewa grammar (indirect-object-like encoding of animate direct objects, agreement restrictions in ditransitives, uniformity of ergative marking, and constraints on incorporation).

The final block of chapters uses functional morphemes to probe the general architecture of the grammar: the derivation of the synthetic/analytic alternation is addressed by Isabel Oltra-Massuet and Ora Matushansky, in close connection with the postsyntactic processes studied by Eulàlia Bonet.

The derivational issues implicit in the choice between a synthetic and an analytic form are the subject of the contribution by Isabel Oltra-Massuet (chapter 1), examining Labovian variability between the three possible realizations of the Catalan past perfective. These realizations consist of a synthetic form, spoken in some varieties of Valencian, Rossellonese, and Balearic Catalan (Majorcan and Ibizan), and two analytic forms, differing in the surface form of the auxiliary historically derived from the present tense of the verb *go*. Deriving all three surface representations from the same underlying structure, Oltra-Massuet accounts for the intraspeaker free choice between the two analytic forms by appealing to a probabilistic application of an extralinguistically conditioned impoverishment rule (cf. Adger and Smith 2005; Nevins and Parrott, 2010).

Ora Matushansky highlights the relevance of the synthetic/analytic alternation at the interfaces in chapter 4, dealing with the derivation of synthetic comparatives and superlatives in English. Arguing against the recent proposals deriving synthetic forms by postsyntactic rules (Embick and Noyer 2001; Embick 2007a), Matushansky advocates returning to Corver's (1997a,b) head-movement analysis. Evidence against treating synthetic forms postsyntactically comes from suppletion (Bobaljik 2012) and coordinated comparatives (Jackendoff 2000), while the effect of scalarity on the availability of synthetic forms and its role in ruling them out with adverbial modification further supports building them in the narrow syntax, rather than postsyntactically.

Syntactic and postsyntactic morphological derivation forms the subject of the contribution by Eulàlia Bonet (chapter 10), investigating the little-known phenomenon of *lazy concord*, illustrated by gender concord in some Spanish dialects and mass concord in Asturian. While in standard Spanish the feminine definite article *la* surfaces as *el* before feminine nouns that start with a stressed [a] (e.g., *el agua* 'the water' instead of **la agua*), in some Spanish dialects the use of the masculine with this class of nouns has generalized to all prenominal adjectives (e.g., **el** *mismo agua* 'the same water', instead of the standard **la** *misma agua*). The hypothesis that gender distinctions may be lost prenominally can then be extended to the Asturian facts, where overt agreement for count-mass distinction but not for gender is neutralized in prenominal adjectives. Bonet argues that a two-step approach to concord, where postnominal concord is done in syntax while prenominal concord is postsyntactic, has advantages over proposals that have been made on parallel "lazy agreement" between the verb and the subject (Ackema and Neeleman 2003; Samek-Lodovici 2002).

As these short summaries show, besides elaborating and advancing the theory of Distributed Morphology, this collection also contributes to the more general body of knowledge in the area of word-level processes and their interaction with syntax, phonology, and semantics: "Morris, Distributed" across the grammar. And yet, while we have all done our best to assemble work here that would develop the theory that we owe to Morris—as his students, colleagues, and friends, we have arranged "Morphemes for Morris Halle" hoping to give him pleasure.

Acknowledgments

The timely completion of this volume would have been impossible had it not been for the incredible dedication of everybody involved. The editors owe a huge debt of gratitude to Isabel Oltra-Massuet, Rolf Noyer, Martha McGinnis, Tatjana Marvin, Heidi Harley and Mercedes Tubino Blanco, Daniel Harbour,

David Embick, Eulàlia Bonet, Jonathan Bobaljik and Susi Wurmbrand, and Karlos Arregi and Andrew Nevins for (mostly!) sticking to the tight deadlines with precision previously unheard of in linguistics, for reviewing each other's papers, for meticulously and repeatedly editing and proofreading, and for being unbelievably supportive of the whole project. We also thank everyone at MIT Press for all their efforts to ensure that the book would be out by July 2013, especially Marc Lowenthal, Marcy Ross, Elizabeth Judd, and Jim Mitchell. Our thanks go as well to Neil Myler for editorial help and for producing the index. Last but not least, we are very grateful to Tim Halle for providing us with the scans of the late Roz Halle's artwork and Morris's photographs.

Note

1. One night, taking a break from this introduction, I called Morris to propose a solution for a problem in Bulgarian morphophonology that we had been working on. Within the first three minutes of the conversation he demonstrated that my solution was wrong; the issues he raised became the basis for the development of the new solution, which will hopefully survive our next conversation [OM].

Abbreviations

#: number   EP: epenthetic

π: person   ERG: ergative

1: first person   EV: Echo Vowel

2: second person   EX: exclusive

3: third person   FEM: feminine

ABS: absolutive F_{TAM}: TAM feature

ACC: accusative   FOC: focus

AGT: agent   FUT: future

ALIGN: epistemically aligned   FUT.IRR: future irrealis

AN: animate   GEN: genitive

ANAPH: anaphoric   GER: gerund

AOR: aorist   GR: greater (GR.PL greater plural)

APPL: applicative   GRP: group

AUTH: author   HEAR: hearer

BEN: benefactive   ICAUS: indirect causative

CAUS: causative   IMP: imperative

CESS: cessative   IMPF: imperfect(ive)

CL: clitic   IN: inclusive

CL.EP: epenthetic clitic   INAN: inanimate

CMPR: comparative grade   INCH: inchoative

COMPL: completive   IND: indicative

COND: conditional   INF: infinitive

CPST: past tense complementizer   INTR: intransitive

DAT: dative   INV: inverse

DEF: definite   IRR: irrealis

DESID: desiderative   LOC: locative

DIM: diminutive   M: Mood

DIR: directive   MAL: malefactive

DL: dual   MASC: masculine

M/M: Morpheme/Morpheme (rule)

M/P: Morphophonological (rule)

MULTISP: multispeaker

NMB: number

NOM: nominative

NONALIGN: not epistemically aligned

NONSTND: nonstandard

OBJ: object

OPT: optative

PART: participant

PASS: passive

PC: paucal

PERF: perfective

PERS: person

PF: perfect

PL: plural

POS: positive grade

PPL: past participle

PR: present

PROG: progressive

PROSP: prospective

PRT: discourse particle

PSN: person

PST: past

Q: interrogative

QUOT: quotative

RED: reduplicative morpheme

REFL: reflexive

REL: relative

REL.FOC: relative focus

SBJV: subjunctive

SG: singular

SIM: simultaneous

SPKR: speaker

SPRL: superlative

STND: standard

SUBJ: subject

TAM: tense/aspect/mood

TEL: telic

TH: theme (vowel)

TOP: topic

TR: transitive

VOC: vocative

1 Variability and Allomorphy in the Morphosyntax of Catalan Past Perfective

Isabel Oltra-Massuet

1.1 Introduction

The past perfective in Catalan illustrates a case of *Labovian variability* (Labov 1969 and related work) in that it shows up to three different forms (1): a synthetic form (S), spoken in some varieties of Valencian, Rossellonese, and Balearic Catalan (Majorcan and Ibizan) (1a); and two analytic forms, the standard (A1), which contains an inflected form historically derived from the present tense of the verb *go* and the infinitive of the corresponding verb (1b); and a nonstandard variant (A2) heard all over the Catalan-speaking area, whose first element resembles the synthetic past (1c). These forms do not express different lexical or truth-conditional semantics, nor do they show different morphosyntactic functions, and individual speakers use some subset of them without distinction.

(1) a.   b.   c.

Synthetic (S)	Standard Analytic (A1)	Nonstandard Analytic (A2)
purificares	vas purificar	vares purificar
purify.2SG.PST.PERF	STND.AUX.2SG purify	NONSTND.AUX.2SG purify
'you purified'	'you purified'	'you purified'

As shown in (2), these data show special morphological features. On the one hand, the nonstandard analytic form A2 is especially interesting in that, at least descriptively, it seems to be built on the standard analytic form A1 with incorporation of the past perfective morpheme *–re–* found in the synthetic form S, and thus being apparently marked twice for past perfective. On the other hand, 1/2PL suppletive forms *anem/aneu* of the lexical verb *anar* 'go' contrast with the absence of suppletion in both auxiliary paradigms.

(2) *Present indicative main verb* go *versus (non)standard past perfective auxiliary* go

	1SG	2SG	3SG	1PL	2PL	3PL	
PrInd lexical *anar* 'go'	vaig	vas	va	anem	aneu	van	
StndPstPerf Aux (A1)	vaig	vas	va	vam	vau	van	+Inf
NonStndPstPerf Aux (A2)	vàreig	vares	va	vàrem	vàreu	varen	+Inf

These data give rise to a number of empirical as well as theoretical questions regarding (i) the synchronic (partial) syncretism between the auxiliary of the A1/A2 forms and the present tense indicative of the irregular lexical verb *anar* 'go' in (2), which in turn brings up the question of the composition of meaning in the analytic past perfective, as well as the issue of the status, representation, and morphophonological realization of roots that seem to compete for insertion into f-morphemes; (ii) the contrast between the allomorphic variation in the 1/2PL present indicative and the paradigm leveling in the past auxiliary; and (iii) the kind—and locus—of structural microvariation that can derive both an analytic and a synthetic phonological realization from the same syntacticosemantic featural content within the grammar of an individual speaker.

Adopting standard syntactic structures and following basic assumptions about the structure of the Catalan verb, I develop a Distributed Morphology account of the above-stated issues that is mainly based on a single functional morpheme MOTION with two phonological exponents, and the interaction between the internal syntax of these forms and the late insertion of underspecified functional Vocabulary items. I further show that intraspeaker variation is subject to morphosyntactic restrictions (synthetic S vs. analytic A1/2), whereas optionality is tied to the probabilistic application of an extralinguistically conditioned Impoverishment rule (standard A1 vs. nonstandard A2).

1.1.1 Catalan Main Verb *Go* vs. Perfective Auxiliary *Go*

Different works have dealt with past perfective in Catalan, such as Badia i Margarit 1951, Colón 1976, Vallduví 1988, Pérez Saldanya 1996, 1998, Comajoan and Pérez Saldanya 2005, Juge 2006, or Jacobs 2011, among others. They are all diachronic approaches mainly hypothesizing about the origin, factors involved, and grammaticalization path from a lexical verb *go* to an allegedly desemanticized past auxiliary,[1] which contrasts with other Romance languages—and most languages in fact, where the verb *go* has been grammaticalized as a future auxiliary. However, I am not aware of any synchronic analysis of the morphology of these forms beyond the classification of *va-* /ba/ as either a past (defective) auxiliary, as in Mascaró 1986, 2002b, Vallduví 1988, Wheeler 1998, Pérez Saldanya 1998, or Gavarró and Laca 2002, or as forms of the present indicative of the verb *anar* with analogic regulation for 1/2nd plural forms, as suggested in Badia i Margarit 1994.[2]

Treating the auxiliary of the analytic past perfective A1 as present indicative forms does not answer the question of the partial syncretism, nor does such a proposal seem able to account for the lack of correspondence between meaning and morphosyntactic content. Likewise, positing a past-tense marker *va-* /ba/ gives an answer to the compositional semantics of these forms, though at the expense of being purely descriptive. In both cases, the fact that a past perfective auxiliary and the present tense of *anar* coincide is a mere accident—that is, a chance homophony.

As for the nonstandard analytic form *vares anar*, almost all descriptive works suggest that it is derived from the synthetic by analogy. In other words, the *-re-* component of the synthetic form *purificares* in (1) appears in the standard auxiliary 'go' *vas* to yield the nonstandard auxiliary *va-re-s*. This, which may be diachronically more or less descriptive, is far from being descriptively adequate and explanatory from a synchronic point of view. As mentioned in Badia i Margarit 1994, 546, and Perea 2002, 641, which are works that acknowledge the analogical status of the segment /r/, the analytic *re*-form is the only construction spoken in some varieties of Catalan; thus, no analogy can be called on from a synchronic point of view.

1.1.1.1 A Brief Note on the Interpretation of Past Perfective

All the forms in (1) have exactly the same semantics, 'you.SG purified', namely simple past perfective. In (3), past perfective contrasts with present perfect, and in (4) with imperfective past.[3]

(3) Ahir   vam-vàrem   fer / férem   coca, però
    yesterday   AUX1-AUX2.1PL   make make-PST.PERF.1PL   cake but
    ja   s'   ha   acabat.
    already REFL   have.3SG.PR   finished
    'Yesterday we baked a cake, but there isn't any more.'

(4) Ahir   vam-vàrem   fer / férem   coca
    yesterday   AUX1-AUX2.1PL   make make-PST.PERF.1PL   cake
    mentre   estudiàvem.
    while   study-PST.PERF-1PL
    'Yesterday we baked a cake while we were studying.'

In Catalan, past perfective differs from present perfect, in that the former can never have any reference to the present, whereas the latter expresses the current relevance of a past situation and cannot have a recent past meaning (as opposed to other Romance languages like some varieties of Spanish), as exemplified in (5)–(6).

(5) Avui / Aquest matí / Aquesta setmana {* vam   fer /
    today   this   morning this   week   AUX.1PL   make
    hem   fet}   coca.
    have.1PL.PR   made   cake
    'Today/This morning/This week we {*baked / have baked} a cake.'

(6) Ahir /   La   setmana passada {vam   fer /   *hem
    yesterday the week   last   AUX.1PL   make   have.1PL.PR
    fet}   coca.
    made   cake
    'Yesterday/Last week we {baked / *have baked} a cake.'

According to Pérez Saldanya (2002), this difference is related to so-called temporal distance between speech-act time and situation time. Whereas the perfect is a hodiernal past, which means that it is confined to the day of the speech act (today), the past is noncurrent or prehodiernal—that is, it is necessarily prior to the day of the speech act.[4]

1.1.2 Intraspeaker Variation and Optionality

The fact that there are three morphologically distinct but semantically equivalent forms appearing in a single paradigmatic cell in a speech community, and that individual speakers freely select some subset of them, raises the question of the existence of doublets (or triplets) in the grammar. It also raises the question of whether they are the result of competing grammars, as suggested for instance in Kroch 1994 or Adger and Smith 2005, or whether they can be the outcome of a single grammar, so that optionality—the probabilistic choice of an item—takes place at the level of use (an external linguistic factor), not at the level of grammar, a desirable result (See Embick 2008 for an extensive discussion of this issue, as well as Nevins and Parrot, 2010.). The kind of structural microvariation that can derive these differences is not trivial. As discussed in Embick 2008, this type of variation involves the nature of optionality in grammar, and the place in the grammar where this optionality is to be found. In the DM model, it follows from basic principles of the theory, that competition is restricted to the level of the morpheme, or in Embick's (2008, 66) words, "There cannot be two outputs for the same input."

In Catalan, the three forms coexist to different degrees in the Valencian, Balearic, and Rossellonese speech communities. Concentrating on the Valencian community, although there is no detailed study of the individual use of these three past forms, data gathered from the personal corpus of Valencian data collected by Esteve Clua (Clua 1999; Viaplana, Lloret, Perea, and Clua 2007)[5] and from my informant point to a crucial difference between morphosyntactic variation and sociolinguistic optionality. On the one hand, some speakers show variation between the S form and the A1/2 forms that is restricted either to certain agreement features, to some conjugation(s), and/or to particular verbs. For instance, some speakers have lost the synthetic form in all singular forms but retain it in the plural (e.g., speakers from Sueca). In other cases, some S singular form is additionally retained for some verbs (e.g., an Alzira speaker who systematically uses the S form with all verbs does not use 1/3SG for the two third-conjugation verbs).[6] There are other examples where variation is affected by the verb itself (e.g., a Sagunt speaker uses the S form for the verbs *cantar* 'sing', *perdre* 'lose', and *servir* 'serve/be useful'). Still other speakers systematically reject the use of the S form in a certain conjugation, the most marked ones (see Oltra-Massuet 2000), which are either third, or second and third conjugations. Despite what may at first glance appear to be random use of these forms, all variation is consistent, in that there do not seem to be cases where the same speaker makes a probabilistic choice between the synthetic variant (S) and the analytic variants (A1/2). Instead, variation between variant S and variants A1/2 is driven by grammatical competition; specifically, variation arises from language-particular restrictions triggered by specific agreement features, conjugations, and/or particular roots that apply in the derivation of the synthetic structure. By contrast, competition between the two analytic variants A1 and A2 is optional and free, because it is driven by sociolinguistic factors, the *re*-analytic being a colloquial nonstandard form in those varieties where it coexists with the standard.[7] Below I will show that whereas the former variation is due to the existence of two competing grammars that give rise to two different syntactic structures, the latter results from the application of an Impoverishment rule that takes place postsyntactically.

1.2 The Analysis of Catalan Past Perfective

The synchronic analysis developed in this section consists of two closely related components. First, we must determine the syntactic structure underlying these past forms, whether it is common to the three forms, and where surface morphosyntactic variation is located. Second, we need to establish the status of the root *anar* 'go' as either lexical or functional, its morphophonological representation, and phonological realization in both environments. A proper understanding of the internal structure of these past forms and root suppletion will account for the (partial) syncretism between the lexical verb and the auxiliary.

1.2.1 Syntactic Structure

For the analysis, I assume standard syntactic structures. In the unmarked case, Catalan chooses to combine Tense (T), Mood (M), and Aspect (Asp) features into a single morpheme for the computational system. In the syntax, a root merges with the verbalizing head v and further undergoes cyclic head-to-head movement all the way up to T (which stands for T/Asp/M or TAM for short). I further assume two language-specific well-formedness conditions. On the one hand, in Catalan all well-formed finite verbs must bear morphological agreement with the subject. This is implemented as Agr adjunction to T in the morphology. At MS, the φ-features of a c-commanding DP in specifier position (the subject) will be copied into this Agr node (see Marantz 1991). On the other hand, I assume Oltra-Massuet's (1999, 2000) proposal for Catalan that at MS, all syntactic functional heads have a theme position adjoined to them.

1.2.1.1 Synthetic past

Given the assumptions just stated, the synthetic past perfective in (7) has the basic syntactic structure in (8), as already suggested in Oltra-Massuet 2000.

(7) *purifiquí, purificares, purificà, purificàrem, purificàreu, purificaren*
    1SG   2SG   3SG   1PL   2PL   3PL
    purify.PST.PERF 'purified'

(8) Synthetic past (after movement—before spell-out)
    [[[√ - v] T]

    Leaving aside 1/3SG, which are special, all the other forms behave as expected for all conjugations. Thus, in the syntax the root first merges with the category-assigning v and undergoes cyclic head-to-head movement up to T, resulting in the structure in (9) after well-formedness conditions and Vocabulary Insertion (VI) (before stress assignment and other specific phonetic rules).

(9) 2SG *Synthetic past* (purificares 'purified.2SG') (*at MS, after Vocabulary Insertion*)
    [√   [v   Th]] [[T   Th] Agr   ]
                            [+PST, +PERF]
      puɾ   ifik   a   ɾ   a   z[8]

1.2.1.2 Analytic Past

For the analytic past, I propose that, as in the unmarked case, TAM features merge into a single node T. However, there is another temporal/aspectual feature, possibly [+ telic] (or [-homogeneous]), also present in the synthetic TAM, that can be realized in a secondary T/Asp node. I assume that a default (neutral) aspectual morpheme can be inserted, namely the infinitive, along the lines of Giorgi and Pianesi's (1998) proposal that the aspectual information is syncretically realized with temporal information and in some cases it is spread through the structure in the form of different T nodes, T1–T2.[9] Adopting proposals in De Swart 1998 for the French *passé simple*, where it is argued that this past perfective is an aspectually sensitive past tense operator that requires a quantized event (i.e., a telic event description as its input), I suggest the basic structure in (10) for the Catalan analytic past perfective. As with the synthetic, the root first merges with the category-assigning *v* and further undergoes movement up to the lower T/Asp head, where I assume that the root meets all its syntactic requirements, as in (10).

(10) Analytic past

$$[_{TP} [\quad T \quad] \ [_{AspP} [\ \sqrt{} \text{ - } v \ ] \quad T/Asp \ ] \ ]$$
$$\text{[+PST, +PERF]} \qquad\qquad\qquad \text{[+TEL]}$$

Support for the existence of such a lower T/Asp head related to telicity (or homogeneity) in Catalan comes from certain aspectual restrictions related to states and individual-level predicates reported in Pérez Saldanya 2002, 2651, which prevents the appearance of such predicates with a past perfective, as illustrated with the ungrammatical examples in (11)–(12), unless they can be coerced into an inchoative predicate, (12b) (see De Swart for this type of semantic coercion).

(11) a. *solgué,   *va   soler
          be.used.to.3SG.PST.PERF   AUX.3SG   be.used.to

  b. *El   llibre   va   tenir vuit   capítols.
      the   book   AUX.3SG   have eight   chapters

  c. *La   taula   on   estudiava   va   ser rodona.
      the   table   where   study.1SG.IMPF AUX.3SG be   round

(12) a. *Als   vint   anys   vaig   ser   alta.
          at.the   twenty   years   AUX.1SG be   tall

  b. Als   vint   anys   vaig   tenir   cotxe.
      at.the   twenty   years   AUX.1SG have   car
      'At the age of twenty I got a car.'

Going back to the structure for the analytic past, in (10), since the root has only raised to AspP, the T features are left stranded without a verbal base that can host them. I propose that a functional auxiliary node v_{AUX} is inserted in the structure to host these T features as a kind of *go*-support, a process parallel to English *do*-support as proposed in Embick and Noyer 2001.[10] The reasons for having *go*-support are discussed in section 1.2.2.1. As proposed for English, this process involves the insertion of a head *v*, a syntacticosemantic object inserted in the syntax to satisfy the locality condition that governs the relationship between T and *v* in (13a), when this condition is not met in the first place, (13b) (Embick and Noyer 2001, 586).

(13) Go-*support*
    a. T must be in an immediately local relationship with *v*.
    b. *v* is syntactically merged onto T when T does not have a *v*P complement.

At MS, the structure must meet all language-particular well-formedness conditions. Since v_{AUX} has been inserted in syntax, a theme position will be adjoined to it. After the application of all well-formedness conditions and VI we obtain the structure in (14) for the 2SG nonstandard analytic form *vares purificar* 'purified.2SG'.[11] See section 1.2.2 for /b/ insertion in v_{AUX}.

(14) 2SG *Nonstandard analytic past* vares purificar[12]

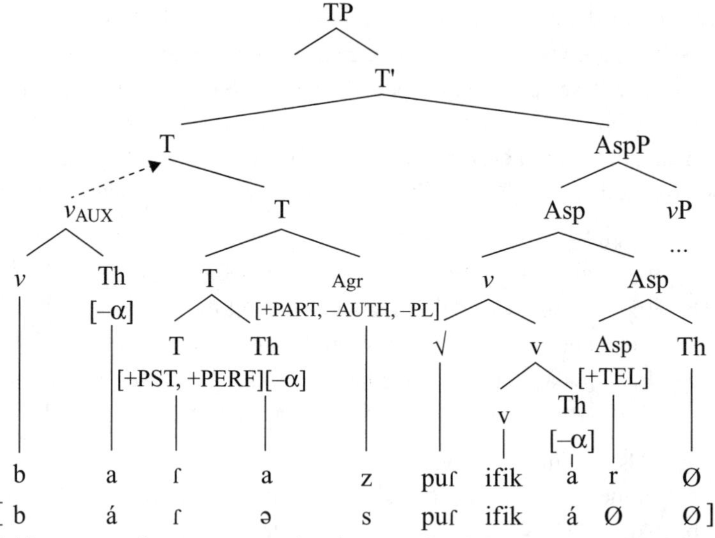

(*After stress placement, vowel reduction, final devoicing, and r-deletion*)

The internal morphosyntactic structure of infinitives, as well as their syntacticosemantic contribution to the structure, is an open issue. Here I assume with Alsina (2002) or Hernanz (1999) that the infinitive Vocabulary Item *-r-* is aspectually neutral in contrast to the participle (perfective) and the gerund (nonperfective or durative), and so it is inserted as a kind of default temporal/ aspectual head. The data in (15)–(16) show that the aspectual interpretation of the infinitive, like its temporal reference, is neutral and must derive from context.

(15) a. En travessar el   carrer (*imperfective aspect*)
          in cross   the street
          'when crossing the street'
     b. Després de   travessar el   carrer (*perfective aspect*)
          after   of   cross   the   street
          'after crossing the street'
          (Alsina 2002, 2392)

(16) a. Quan arribi,   penso   comentar- li   el
          when arrive.1SG.PR.SBJV think.1SG.PR.IND comment 3SG.DAT the
             que
             what
          m'   has   dit. ('li   ho comentaré')
          1SG.DAT have.1SG.PR told (3SG.DAT it   comment.1SG.FUT)
          'When she arrives, I intend to tell her what you told me.' ('I will
             tell her' = future)
     b. El   vaig   sentir arribar a la una
          3SG.ACC AUX.1SG hear   arrive at the one
          ('com arribava a la una')
          'I heard her arrive at one.' ('that she arrived at one' = past)
          (how   arrive.3SG.IMPF at the one)
          (Pérez Saldanya 2002, 2651)

Turning to the analysis of the standard analytic *vas purificar*, this form completely lacks the T node [T-Th], which in the nonstandard analytic is realized by the string /ɾe/. This seems a clear case of impoverishment. Specifically, in the morphology a T node specified as [+PST, +PERF] is impoverished, as stated in (17). Everything else will apply as in the *re*-analytic variant.[13]

(17) *Impoverishment Rule*
     $[_{\mathrm{T}}$ +PST, +PERF$] \rightarrow \emptyset$

Note that both analytic forms are interpreted in the same way at LF, because they have exactly the same syntacticosemantic structure. They are distinguished at MS, as a result of a language-particular variable rule whose extralinguistic (sociolinguistic) context of application will be specified with a probabilistic index (Adger and Smith 2005; Nevins and Parrot, 2010)—that is, this rule applies in a nondeterministic manner. Since (17) is a specific rule subject to external non-linguistic factors, I assume that it will apply after the general, deterministically applying well-formedness conditions, so that the node Agr will first be adjoined to T. When the rule in (17) applies, it deletes the terminal node containing the features that trigger the rule. However, it leaves the higher T that hosts the Agr features, which is necessary for stress assignment (Oltra-Massuet 1999, 2000).

In speakers showing variation between a synthetic and an analytic form, choice of one form or the other is triggered by certain morphosyntactic features (agreement and conjugation class) and/or specific lexical roots.

1.2.1.3 Synthetic vs. Analytic Past: One or Two Structures?

As I have already mentioned, in DM competition can only occur at the level of VI. This means that for each input structure there will just be one output structure. This in turn means that there should be a different structure for each Catalan past form—that is, there should be three different morphosyntactic structures. At the same time, these three structures must convey exactly the same semantic meaning; in other words, they must contain exactly the same semantic features that receive the same semantic interpretation. However, if we assume a direct relation between syntax and semantics, as suggested in Embick 2007a, so that sameness of meaning corresponds to identity of syntactic structure, we should have the same underlying syntactic structure for all forms, surface differences arising from operations triggered by some marked feature at MS. Note that the fact that we have variation between a word and a complex phrase is irrelevant in DM. As shown in Embick's (2007a) analysis of the synthetic and analytic comparative in English,[14] there is no word-phrase competition, but a structural difference that involves the way syntactic heads are put together or "packaged" for phonological interpretation at MS (see also Embick and Marantz 2008).

Nonetheless, I have proposed two structures, one for the synthetic and a second one to derive the two analytic forms, where the same semanticosyntactic features are merged in different terminal nodes even though they receive exactly the same interpretation. According to Embick (2008, 65), "If there is one input $\mathcal{N}$ to a syntactic derivation, and we find two distinct forms derived from this $\mathcal{N}$, then there must be distinct grammars at play." Since I have proposed two output syntactic structures for the same input for the past perfective forms, this means that variation between the synthetic and the analytic must be the result of different competing grammars. In this section, I briefly discuss some empirical and theoretical problems that the assumption of a single structure would pose.

I have suggested that the synthetic past perfective is unmarked, like all simple tenses except future and conditional, so that TAM features are merged into a single node, whereas the analytic is marked and requires two temporal/aspectual heads. We could assume a splitting of TAM into $TAM_{[+PST,+PERF]}$ and $T/Asp_{[+TEL]}$ in both analytic and synthetic forms, as in (18).

(18) Underlying structure for past perfective (simplified, before movement)
$[_{TP}$ TAM $[_{AspP}$ T/Asp $[_{vP}$ v √]]]
[+PST, +PERF]   [+TEL]

However, assuming movement of v to T in the synthetic form would give us the wrong derivation in (19), *purificàrres*, so that some additional machinery such as a degemination readjustment rule should be stipulated.

(19) Ungrammatical synthetic past (*purificàrres 'purified-2SG')
[√ [v Th]] [**T/Asp Th**] [**TAM Th**] Agr]
               [+TEL]   **[+PST, +PERF]**
puɾ   ifik a   ɾ   **Ø**   ɾ   **a**   z

To derive the analytic form, raising of v to TAM should be blocked in the syntax due to some language-specific constraint, either by means of the presence of a specially marked root or by certain morphosyntactic features present in a DP subject. This would entail that roots with this kind of restrictive information should be present in the syntactic derivation. Although the early insertion of roots has already been defended in the DM literature (see e.g. Embick 2000; Siddiqi 2009), more research is needed to evaluate the different predictions of these two proposals, since both involve language-specific variation in syntax triggered by specific lexical roots or certain diacritic features of roots.

Note that I have disregarded an analysis where the presence of a marked feature [+PERF] triggers the split of the AspP from T, as was proposed for a [+FUT] Mood in the future and conditional in Oltra-Massuet 1999, 2000. This would give us the wrong synthetic form in (20), *purificates*.[15]

(20) *Ungrammatical synthetic past* (*purificates 'purified.2SG')
[√ [v Th]] [**Asp Th**] [**T Th**] Agr]
               [+PERF]   **[+PST]**
puɾ ifik a   **t**   **Ø**   **Ø** **a**   z

Recall that the infinitive Vocabulary Item is temporally and aspectually neutral in contrast to the participle (perfective) and the gerund (nonperfective or durative); thus, having a [+PERF] Asp head in the analytic form would imply insertion of the participle Vocabulary Item /t/ instead of the neutral infinitive. Note that in both structures, we require a [+PERF] aspectual realized in T. It does not seem adequate to propose a splitting of TAM triggered by a feature [+PERF] that needs additional operations to end up being realized back in T.

An anonymous reviewer suggests that the simplest analysis would be to analyze both forms, S and A1/2, as derived from a single underlying structure without default merging of Tense and Aspect. Optional movement of the complex [Asp—v—√] to T would yield the synthetic; otherwise an auxiliary would be inserted in the analytic as proposed above. This proposal would preserve the uniformity of syntactic structure. However, it would also block the default merging of Tense, Mood, and Aspect features—which would then be merged again in the synthetic, a process that applies across the board in the language in all tenses but Future and Conditional. In addition, such an analysis would also require a unique degemination readjustment rule and would pose the same problems discussed so far. On the one hand, syntactic movement to T in the synthetic configuration would have to be determined by language-specific features of lexical items; on the other hand, having an aspect terminal node specified as [+PERF] would result in the insertion of a participle Vocabulary Item /t/ instead of the neutral infinitive Vocabulary item /ɾ/.

1.2.2 Root Suppletion

Turning to the status of *va-* as a suppletive form of a lexical verb *anar* 'go', this is problematic since in DM suppletion is restricted to the functional vocabulary of the language that competes for insertion, for instance auxiliaries or light verbs such as *go* (e.g., Halle and Marantz 1993; Marantz 1995, 1997; Embick and Halle 2005; Embick and Marantz 2008; Embick 2010).[16] Lexical roots showing allomorphy are derived through readjustment rules; if allomorphs are phonologically unrelated, the system treats them as contextually determined exponents of an f-morpheme that compete for insertion in a deterministic manner. This means that *anar* must be an f-morpheme.

1.2.2.1 *Lexical and Functional* Anar *'Go'*

The verb *anar* 'go' is morphologically unique in Catalan, in that it shows the palatalization typical of second-conjugation verbs (*veure–veig* 'see–I see', *fer–faig* 'do–I do'), and the root is followed by the default theme vowel /i/ of third-conjugation verbs in the future and the conditional, but it behaves as a first-conjugation verb in all other forms (Mascaró 1986, 121), as shown in (21).

(21) *Lexical verb* anar '*go*' (anat '*gone*'—anant '*going*')

	PRIND	PRSBJV	IMP	IMPFPST	PERFPST	IMPFSBJV	FUT	COND
1SG	vaig	vagi		anava	aní	anés	aniré	aniria
2SG	vas	vagis	vés	anaves	anares	anessis	aniràs	aniries
3SG	va	vagi	vagi	anava	anà	anés	anirà	aniria
1PL	anem	anem	anem	anàvem	anàrem	anéssim	anirem	aniríem
2PL	aneu	aneu	aneu	anàveu	anàreu	anéssiu	anireu	aniríeu
3PL	van	vagin	vagin	anaven	anaren	anessin	aniran	anirien

As discussed in Pomino and Remberger 2008 for Spanish *ir*, suppletion becomes especially interesting—and problematic—in cases where we have a single item that behaves sometimes as a functional element, sometimes as a lexical unit. This is the case of the verb *anar* 'go'. This verb appears in different contexts, auxiliary or lexical, as illustrated in (22).

(22) a. En Joan va   fer deures.
        the Joan AUX.3SG do homework
        'Joan did homework.'

  b. En Joan va   fent   deures.
       the Joan go.3SG.PR.IND do.GER homework
       'Joan keeps doing homework.'

  c. En Joan va   a fer deures.
       the Joan go.3SG.PRS.IND to do homework
       'Joan is going right now to do homework.'

  d. En Joan va   a Tarragona.
       the Joan go.3SG.PRS.IND to Tarragona
       'Joan goes to Tarragona.'

  e. En Joan va   en cotxe / de pressa / despullat.
       The Joan go.3SG.PRS.IND in car   of hurry   naked
       'Joan goes by car / fast / naked.'

Although a full account of the properties of *anar* 'go' is beyond the scope of this chapter, I will provide the basic insight that allows the treatment of all instances of *anar* as derived from a single abstract f-morpheme. I propose that this f-morpheme contributes a meaning of unspecified or neutral MOTION— that is, it is a v_{MOTION}. The exact semantics of this event head is derived from the syntactic environment for each particular instance. When the v_{MOTION} formative merges in the syntactic derivation in its regular lower v location, it functions as the regular verb *go*, as in (22c–e). In this case, it is interpreted as temporal motion (immediate prospective) or spatial motion (directional, in a certain manner) depending on the kind of complements it takes—for example, complements denoting prospectivity (see Pomino and Remberger 2008) or immediate future (22c),[17] a path (22d), or a manner (22e). In addition, it can also be inserted higher in the construction, as in (22a), and presumably also (22b), if we assume that the latter is a case of the auxiliary *go*. In these cases, v_{MOTION} will be inserted postsyntactically, at the level of MS. That is, in the syntax, a v_{AUX} is inserted in the manner shown in section 1.2.1.2, but it is in the morphological component that a v_{MOTION} is inserted as an unspecified or default auxiliary. In the case of the past auxiliary, it is interpreted as past tense, because T is past and it is so interpreted at LF. This is not a stipulation; it just follows from the fact that a default v_{AUX} merges with a perfective past temporal head.

Let us make the insertion of the syntacticosemantic formative v_{MOTION} more explicit. The idea is that before VI applies to a v_{AUX} sister of a past perfective T, the auxiliary Vocabulary items in (23) compete for insertion,[18] and the default f-morpheme MOTION is selected, which then will refer to its corresponding set of phonological exponents for VI, where again the default exponent is selected. In other words, the rule of *Go*-support in (13) inserts a neutral v_{AUX} category as a terminal node, but it does not specify its subtype or "flavor." Later, at MS and before Vocabulary Insertion, a second process adds a MOTION feature as an elsewhere option.

(23) *Auxiliaries that compete for insertion into v_{AUX} in Catalan*
    HAVER ↔ [+AUX, +PERF] 'have'
    SER   ↔ [+AUX, +PASS] 'be'
    ESTAR ↔ [+AUX, +PROG] 'be-stative'
    DESPLAÇAMENT ↔ <elsewhere> 'motion'

This proposal is similar to Embick's (1997) *dissociated morpheme*, but with *dissociated features* instead.[19] That is, a *v* terminal node is added to the structure by the *Go*-support rule and produces an auxiliary verb structure v_{AUX}-T in syntax. At this point, v_{AUX} is fully content-neutral, and it functions as a syntactic well-formedness item, as stated in section 1.2.1.2. At MS, that terminal node is filled with features that determine which actual vocabulary items will be able to spell out the terminal node in a process of default dissociated feature insertion of MOTION that applies before Vocabulary Insertion. Apart from the set of typical auxiliary verbs *haver*, *ser*, *estar*, no other Catalan verb shows the kind of suppletive allomorphy found in *anar*, which forces the system to treat it as an f-morpheme. My contention is that being verbal, which means being categorized as *v* in DM, makes this f-morpheme suitable to compete for insertion into a *v* node, and since it is completely underspecified, v_{MOTION} is selected as a default f-morpheme in the past perfective, thus acquiring the apparent surface status of a specific past marker. After v_{MOTION} is selected, VI applies.

The general assumption of a separate auxiliary past form implies the existence of two unrelated *go* elements, an auxiliary and a lexical verb (e.g., see Pomino and Remberger 2008 for a formal account). Otherwise, one is forced to claim that a new kind of element originates when lexical categories are merged as functional heads. Cardinaletti and Giusti (2001) argue for a semi-lexical motion verb in Italian, where the auxiliary loses all of its selectional and semantic properties and retains only its morphological properties. More recently, it has been explicitly claimed that the Catalan *go*-past auxiliary "is void of semantic content" (Jacobs 2011, 228, and references there). Despite the assumption of a basic semantics of undefined motion for an abstract f-morpheme MOTION, note that the *Go*-support analysis in section 1.2.1.2 solves a long-standing debate about the synchronic relation between the lexical verb *anar* and the auxiliary *go*-past. In (22a–e) we are dealing with just one form, whose complex interpretation derives from its syntactic environment. No claims about new categories or about the multiplication of *go* elements are necessary.

1.2.2.2   Root Suppletion as Contextual Allomorphy of F-morphemes

I have already shown that the two instances of *va-* are syntactically related; both derive from an abstract f-morpheme MOTION that can be inserted at different structural positions at different stages of the derivation, which rules out the possibility of a chance homophony. That also means that /b/ and /an/ are phonological exponents of this f-morpheme MOTION. The next thing to do is to determine their contexts of insertion. When considering the entire paradigm in (21), at first glance it would appear that /b/ is the specific, being inserted in a [-PST] context, and /an/ would be the elsewhere exponent. Following standard elsewhere ordering principles within the DM framework (Halle and Marantz 1993; Noyer 1992), contextually conditioned VI blocks the insertion of the expected Vocabulary item. And, in principle, we would expect /an/ in the whole paradigm, so that we would have the Vocabulary Items in (24), where the auxiliary appears as a homophonous unrelated Vocabulary Item.

(24) *Vocabulary items for f-morpheme v_{MOTION}—Option 1*

$$v_{\text{MOTION}} \leftrightarrow \text{/b/} \begin{Bmatrix} /[_{\text{AUX}}\underline{\quad\quad}] \\ /\underline{\quad\quad}[-\text{PST}] \end{Bmatrix}$$
$$\text{/an/} \begin{Bmatrix} /[+\text{PART}, +\text{PL}] \\ /<\text{elsewhere}> \end{Bmatrix}$$

On the other hand, Bobaljik (2002) views cases of true syncretism as neutralizations where a less marked Vocabulary Item is inserted. It has been claimed that the past auxiliary in Catalan underwent a process of phonological regularization during the grammaticalization process, which I see as a case of neutralization, where the elsewhere exponent has been inserted. I propose the Vocabulary Items in (25).

(25) *Vocabulary items for f-morpheme v_{MOTION}—Option 2*

$$v_{\text{MOTION}} \leftrightarrow \text{/an/} \;/\; \underline{\quad\quad} [+\text{F}] \quad ([+\text{F}] = \text{marked context})$$
$$\text{/b/} \;/\; <\text{elsewhere}>$$

These Vocabulary Items are inserted where expected, assuming that the context [+PART, +PL] of 1/2PL present indicative forms is a marked context, which is unproblematic (e.g., Siewierska 2004). In the auxiliary, which hosts the very marked T features [+PST, +PERF], we do not obtain the expected results, though. There are different options to formally obtain paradigm leveling in the auxiliary.[20] Paradigm leveling could be related to having the operation of VI taking place in different phases, so that the marked features of T would not affect the insertion of /b/ in v_{AUX}.[21] Another possibility is that when a completely empty syntactic object is filled with an elsewhere v_{MOTION} f-morpheme, the operation inhibits the visibility of contextual features (see Calabrese 2003 for the notion of inhibition). This may also be related to a more general theory of default exponence and its interaction with markedness, a project that will have to await further research.[22]

1.2.3 Syncretism

At this point, we can explain the surface similarity between the present indicative of *anar* and the auxiliary of the standard analytic past. In Oltra-Massuet 1999, 2000, I proposed the syntactic structure in (26) for the present tense, which can now be compared to the structure I have proposed for the auxiliary of the analytic past in (27). In the forms other than 1/2PL, the syncretism arises despite the different morphosyntactic structure. Vocabulary Items, since they are underspecified, may be inserted in different morphosyntactic positions. For instance, in (26) whereas the vowel /a/ in the present indicative is the theme vowel realized in the theme position adjoined to a verbalizing head *v*, the vowel /a/ corresponds to the one adjoined to the auxiliary head v_{AUX} in the past perfective in (27). Lack of syncretism in 1/2PL is due to default insertion and paradigm leveling in the context of a v_{AUX}.

(26) 2SG *Present indicative* vas

    [[v_{MOTION} **Th**]][T/Agr]]

        b   *á*   *z*

(27) 2SG *Analytic past perfective* vas

    [[v_{AUX} **Th**] Agr]

        b   *á*   *z*

1.3 Conclusions

The DM analysis of the various past perfective forms in Catalan developed in this chapter shows that the late insertion of functional material together with the syntactic structures proposed explains a surface morphosyntactic mismatch, and accounts for the partial syncretism between lexical and functional *anar*. That *va-* /ba/ contributes past temporal meaning whereas the infinitive provides the lexical content to the periphrastic forms now receives an explanation without stipulating the existence of a semantically and morphosyntactically unrelated past auxiliary. The proposal of an abstract MOTION f-morpheme solves a long-standing debate on the exotic status of a *go*-past auxiliary in Catalan, which, I claim, can account for all instances of *anar*. More generally, it offers a new approach to the status of lexical roots that compete for insertion into f-morphemes that could be extended to other languages where *go* and similar roots have both lexical and functional uses.

This proposal can be related to the main hypothesis about the diachronic path from lexical verb *anar* 'go' to auxiliary *va-* put forward by Pérez Saldanya (1996, 1998). Although a more detailed analysis is necessary, note that the realization of *va-* as a kind of auxiliary support at the level of T, and an infinitive realizing a telicity-related temporal/aspectual head, is in accordance with the final stage proposed in Pérez Saldanya's account of the historical path from a lexical verb *anar* 'go' to a tense marker through an intermediate stage as an aspectual marker.

Notes

I am especially indebted to two anonymous reviewers for valuable comments and suggestions that have improved the chapter substantially. Thanks to E. Bonet and O. Matushansky for very insightful suggestions, even though some of them will have to await future research, as well as to J. Mascaró for helpful comments to a previous version, to E. Clua for his Valencian corpus, to G. Gómez Duran for data on Rossellonese, and to A. Guardiola Severí for Valencian. All remaining errors are my own. This work has been funded by MICINN grants FFI2010-22181-C03-01 and FFI2009-07114/FILO.

1. See for instance Jacobs 2011 and the references there for the idea that the auxiliary contains no semantic notion of motion.

2. But see Badia i Margarit's (1981, 371) claim that Modern Alguerese uses the forms in (i).

(1) {*vaig   vas   va   anam   anats   van*} +   *cantar*
     AUX.1SG AUX.2SG AUX.3SG AUX.1PL AUX.2PL AUX.3Pl sing
     'sang'

Note that these forms exhibit a closer relationship between the past auxiliary of the analytic form and the lexical verb *anar* 'go'. Indeed, these forms would correspond to the *Stage 0* hypothesized by Pérez Saldanya (1998) for the grammaticalization path to the past auxiliary *va-*, before phonological reduction (i.e., paradigm leveling) eliminated the stressed forms *anam, anats* (see also Colón 1976).

3. See Cipria and Roberts 2001 and De Swart 1998 for the semantic differences between perfective and imperfective in other Romance languages (Italian and French, respectively).

4. See Pérez Saldanya 2002 for aspectual differences between the perfect and the simple past. See more generally Iatridou, Anagnostopoulou, and Izvorski 2001 for further differences.

5. Clua's (1999) corpus contains verbal data from 70 native speakers of Valencian from 22 different towns for 5 verbs: 1st conjugation *cantar* 'sing', 2nd conjugation *perdre* 'lose' and *beure* 'drink', and 3rd conjugation *sentir* 'feel/hear' and *servir* 'serve/be useful'. Nineteen speakers from seven areas (Alzira, Elx, Guardamar, Llíria, Sagunt, Sueca, València) use the S variant, often combined with A forms.

6. It has generally been observed that the 1SG synthetic form has been progressively replaced by the analytic, so that most speakers always use an analytic 1SG (see e.g. Pérez Saldanya 2002 on such sociolinguistic variation). First and third person singular are highly idiosyncratic.

7. It should be added that the 1SG nonstandard analytic A2 form *vàreig* is not used at all by the Valencian speakers of the corpus, though it can be heard in other varieties that use A2.

8. Mascaró (1986) argues that the phonological exponent for 2SG agreement features is an underlying /z/ that turns into [s] by means of a general phonetic process of final devoicing.

9. Note, however, that Giorgi and Pianesi (1998) propose a single T node for the simple past.

10. See Pomino and Remberger 2007 for an apparently similar proposal to account for Spanish *ir* 'go'. In their analysis, a *go*-support is inserted in the context of a root of a lexical verb *ir* that has no phonological content. My proposal differs greatly, since v_{AUX} is adjoined to a functional node T that has no verbal base, exactly as in English *do*-support.

11. In principle, there is no syntactic evidence for having the auxiliary head merged in the syntax. It could as well be inserted in the morphology, along the lines proposed by Ippolito (1999), based on Giorgi and Pianesi 1998, where she suggests that higher Is (or Ts in Giorgi and Pianesi 1998) do not need to be assigned a V (i.e., an event position) in order to be visible to the computation. Ippolito (1999, 119) further proposes for Italian that it "is a language-specific well-formedness condition that each I require a V. Therefore, insertion of AUX happens at MS, exactly like insertion of Agr."

12. I have assumed that a theme position is adjoined to the lower T/Asp head as well, spelled-out as zero, which is in fact the default theme of the nominal environment in Catalan. The status of this head is controversial. In Oltra-Massuet 1999, 2000, I assumed that the infinitival morphosyntactic structure contains a nonfinite defective T, hence not subject to the well-formedness condition on functional heads F, whereas in Oltra-Massuet and Arregi 2005, we assumed that the infinitive in Spanish realizes some nonverbal functional head. Given the well-known dual or hybrid nature, verbal and nominal, of the infinitive (Hernanz 1999), I leave the exact details of this derivation for further research.

13. Note, however, that in certain nonstandard varieties of Catalan we get *vem-veu purificar* /b′em-b′ew puɾifikál/ instead of the regular *vam-vau purificar* /bám-báw puɾifikál/ in the 1/2PL forms. Because of space limitations, I cannot discuss the details of this variation, though note that this could be taken as morphosyntactic evidence for having an auxiliary head merged in syntax, as opposed to those varieties where we always have *va-* /ba/ (see note 11).

14. See Matushansky (chapter 4, this volume) for an analysis of synthetic comparatives and superlatives that differs from Embick 2007a.

15. Note that if the phonological exponent of the participle vocabulary item /t/ was blocked for some reason, we still would get a wrong form *purificaraes* 'purified.2SG'. Appeal to some rule of vowel reduction to derive the correct *purificares* would not be possible either, since, according to Mascaró (2002a, 110), some speakers would allow such a pronunciation in words like [sáəɾə] for Sàhara.

16. But see Bonet and Harbour 2012, where it is argued that root allomorphy does exist and cannot be restricted to f-morphemes. Still, these authors agree that auxiliaries and light verbs like *be* and *go* are part of the language's functional vocabulary. See also Harley 2011 for the claim that true root suppletion is a real phenomenon.

17. Note the contrast between (22b) and the ungrammatical Catalan example in (i). Catalan does not have a future auxiliary *go*, as opposed to most other Romance languages, such as Spanish, as exemplified in (ii). (22b) is an example of the lexical verb *go* in Catalan, which is interpreted as immediate future.

(i) *Demà   vaig   a fer   una coca.
     tomorrow go.1SG.PRS.IND to make   a   cake

(ii) Mañana   voy   a hornear un   bizcocho.
     tomorrow go.1SG.PRS.IND   to bake   a   cake
     'Tomorrow I am going to bake a cake.'

18. The specification of these Vocabulary items is just illustrative; different and/or additional features may be necessary.

19. I am grateful to an anonymous reviewer for helping me clarify and find the right way to express the idea I had in mind with respect to the process of insertion of MOTION.

20. A reviewer suggests having a marked /b/ Vocabulary item specified for insertion into a position structurally adjacent to T. Since 1/2PL forms would have undergone a process of T deletion, the elsewhere /an/ would be correctly inserted (see Oltra-Massuet 1999, 2000). However, this possibility would also rule out /b/ insertion in the non-standard A2 *vares* paradigm, for which I have proposed the T Impoverishment rule in (17) above—apart from other structural changes.

21. See Bobaljik and Wurmbrand (chapter 11, this volume) for the relationship between syntactic phase and morphological cycle in the analysis of suppletion in English comparatives and superlatives.

22. The suppletion contrast between 1/2PL and the rest of the forms in the present tenses (indicative and subjunctive) is a pervasive phenomenon in the Romance languages and it has received ample attention. It has generally been analyzed as a case of phonologically governed suppletion—for instance, with Maiden's (1992, 2004b) Romance N-pattern based on stress, or more recently Anderson's (2008) analysis of Surmiran, a Swiss Rumantsch variety, among many others. For reasons of space I cannot discuss this very complex issue here, which is a topic in its own right. Let me just mention that, whereas emphasis has been placed on deriving this suppletion pattern on the basis of underlying stress facts and the presence of the theme vowel, it may be worth approaching it from a nonphonologically oriented perspective, along the lines of Oltra-Massuet 1999, 2000, as well as Oltra-Massuet and Arregi 2005, where the predictive nature of stress in Catalan and Spanish verbs is shown to derive from a morphosyntactic analysis of the verbal forms.

2 Phonological and Morphological Interaction in Proto-Indo-European Accentuation

Rolf Noyer

The reconstruction of the grammar of word accentuation in Proto-Indo-European (PIE) has long been a central topic in historical linguistics, as well as the focus of a number of studies within generative phonology of the daughter languages that preserve relicts of the anterior system (Halle and Kiparsky 1977; Halle and Vergnaud 1987). Within traditional historical linguistics a particular reconstruction of PIE accentuation, based principally on the work of Warren Cowgill, Jochem Schindler, and Helmut Rix from the mid-1970s onward (e.g., Schindler 1972, 1975a, 1975b; Rix et al. 2001), has been accepted as a standard working hypothesis in much current research. Evidence for this reconstruction comes partly from the accentual systems of certain attested Indo-European languages (Vedic Sanskrit, Russian, Lithuanian, Greek); reflexes of the pervasive pattern of vowel alternations (traditionally known as *grades*), which was partly correlated with accent position, especially in Anatolian, Greek, and Indo-Iranian; and the effects of accent position on obstruent voicing in Proto-Germanic (Verner's law) and on pitch-accent types in the Baltic languages. The Schindler-Rix reconstruction is the basis of the analysis to be presented here; in particular I will follow the recent detailed presentation of Ringe 2006, from which the majority of the data has been taken; see also Fortson 2010 for a basic overview of the reconstruction of PIE and its accentuation.

Halle (1997b) presents a preliminary analysis of the Schindler-Rix reconstruction within Simplified Bracketed Grid Theory (SBGT, Idsardi 1992). The present chapter is intended to supplement this line of research and to widen the scope of discussion to include the role of cyclic and noncyclic phonology in producing the accent classes of derived stems in PIE.

In addition, I depart from earlier studies in a number of details. First, although "fixed" accent ("acrostatic" and "mesostatic") stems emerge as a pervasive class in later languages such as Russian and Lithuanian, they are understood here to be relatively marked in comparison with the older stem type in which the position of accent vacillates between two adjacent syllables ("hysterokinetic" and "proterokinetic"). Second, the underlyingly prespecified metrical structure of PIE morphemes includes not only left boundaries of metrical constituents ("feet"), as in Halle 1997b, but also right-constituent boundaries, which produce a new type of "back-accenting" behavior. Third, following ideas introduced in Halle 1998 in the analysis of English stress, I make crucial use of the premise that certain morphemes may fail to project metrical grid marks, and thus be invisible to metrical computations; this invisibility is distinct from ordinary "extrametricality" in which peripheral elements remain external to metrical constituents, although still metrically visible.

2.1 Halle and Vergnaud's Basic Accentuation Principle

In their surface form, words in Proto-Indo-European (PIE) are hypothesized to have contained a single accented syllable, where accent was phonetically implemented as a high tone—for example, *$d^h ugh_2 t\acute{e}r$ 'daughter', *$\acute{e}\hat{k}u\!\!\:os$ 'horse', *$snus\acute{o}s$ 'daughter-in-law', *$s\acute{o}h_2u\!\!\:l$ 'sun'. Although the position of accent in a word is often described as "free"—as if entirely an arbitrary property of a word's inflectional "paradigm"—it is in fact derivable from a combination of the underlying properties of the word's root and affixes and a system of general phonological rules.

Halle and Vergnaud 1987, and later Halle 1997b, propose a Basic Accentuation Principle (BAP) for PIE, based on an analysis of Russian, Vedic Sanskrit, and Lithuanian, which are argued to preserve the essential character of the historically anterior system.

(1) *Basic Accentuation Principle*
   a. In underlying representation, stems and affixes are either accented or unaccented.
   b. Leftmost: Stress falls on the leftmost accented syllable if any.
   c. Default initial: Otherwise stress falls on the initial syllable.

In the modern Russian words in (2), for example, an underlyingly unaccented stem such as *skovorod-* 'frying pan' contrasts with an underlyingly accented stem such as *kómnat-* 'room' when an accented suffix such as NOM.SG -*á* is added (underlying accent is indicated by underlining):

(2) a. skovorod-a̱ / skóvorod-y   'frying pan' NOM.SG/GEN.SG
   b. ko̱mnat-a̱ / ko̱mnat-y   'room' NOM.SG/GEN.SG

In *skovorodá* the suffix is the leftmost underlying accent (it is the only one), and thus has surface accent. In the GEN.SG *skóvorody,* however, neither the stem nor the suffix *-y* has an underlying accent; in such cases accent falls by default on the initial syllable of the word. In contrast, in *kómnat-* surface accent remains fixed on the initial syllable irrespective of the accentuation of the suffix, since the underlying accent of the stem will always be to the left of any suffixal accent.

2.2   Proto-Indo-European Accent Patterns

Certain aspects of the Russian pattern reviewed above have clear parallels in PIE. As in Russian, certain PIE case-number endings (traditionally called *desinences*) attract stress whereas others do not. The unaccented desinences occur in the direct cases (nominative, vocative, and accusative) and the accented desinences in the oblique cases (instrumental, dative, ablative, genitive, and locative).

In addition, typically but not always, unaccented syllables reduce to "zero grade"—that is, if they contain a nonhigh vowel *e o a* it will be deleted. Alternations between *o* ~ *e* and *ē* ~ *e* are also observed in various contexts.

The examples in (3) contrast the behavior of an accented stem *$\!nók^wt$-/*$\!nék^wt$-* 'night' with that of an unaccented stem *$\!léi$-mon-/*$\!li$-mn-* 'lake'. (Note that the locative desinence, which at least in early PIE is -Ø, behaves like an accented desinence, as opposed to the vocative -Ø, which is unaccented.)

(3) SG.NOM   \quad *nók^{w}t-s   \qquad *léi-mon-s (→ *léimō)
      VOC   \qquad *nók^{w}t-Ø   \qquad *léi-mon-Ø
      ACC   \qquad *nók^{w}t-m̥   \qquad *léi-mon-m̥
      DAT   \qquad *nék^{w}t-ei   \qquad *li-mn-éi
      LOC   \qquad *nék^{w}t-Ø   \qquad *li-mén-Ø
      DUAL DIRECT   *nók^{w}t-h₁e   \qquad *léi-mon-h₁e
      PL.NOM/VOC   *nók^{w}t-es   \qquad *léi-mon-es
      LOC   \qquad *nék^{w}t-su   \qquad *li-mn̥-sú

In the traditional nomenclature of PIE accentology, an accented stem such as *$\!nók^wt$-/*$\!nék^wt$-,* with accent fixed on the initial syllable, is called *acrostatic*, whereas an unaccented stem such as *$\!lei$-mon-/*$\!li$-mn-* is called *amphikinetic*. The locative singular forms in -Ø, although apparently exceptional, are not: if surface accent is generated on the zero desinence it is automatically retracted to the preceding syllable. It should be clear, then, that the acrostatic and amphikinetic patterns conform to the BAP and require no further discussion.

In addition to these two accent-ablaut patterns, however, the Schindler-Rix system recognizes three other basic types: mesostatic, hysterokinetic, and proterokinetic. These patterns are conventionally schematized using a morphological analysis, due originally to Benveniste, that divides the word into a Root, a Suffix, and an Ending (or Desinence): $R + S + E$:

(4)

		Unaccented ending	Accented ending	Null LOC.SG	Example stem
a.	Acrostatic 'night'	Ŕ+S+E	Ŕ+S+E	Ŕ+S+Ø	*$nók^wt$-/*$nék^wt$-
b.	Amphikinetic 'lake'	Ŕ+S+E	R+S+É	R+Ś+Ø	*$léi$-mon-/*li-mn-
c.	Mesostatic 'daughter-in-law'	R+Ś+E	R+Ś+E	R+Ś+Ø	*$snus$-$ó$-
d.	Hysterokinetic 'daughter'	R+Ś+E	R+S+É	R+Ś+Ø	*d^hug-$h_2tér$/-h_2tr-
e.	Proterokinetic 'thought'	Ŕ+S+E	R+Ś+E	R+Ś+Ø	*$mén$-ti-/*$mn̥$-$téi$-

The mesostatic pattern has fixed accent on the suffix; this leads naturally to an analysis where in such stems the root has no underlying accent but the suffix does, and thus surface accent remains fixed on the suffix regardless of the desinence. The hysterokinetic and proterokinetic types, however, elude an immediate analysis within the terms presented thus far.

Although various stipulations could in principle be adopted to generate these two additional patterns, the approach taken here aims to meet the following criteria of explanatory adequacy. First, whatever grammar is reconstructed for PIE should not be sui generis. That is, a reconstructed PIE grammar should be no different in principle from a grammar of a currently spoken language;[1] more specifically, its system of accentuation should be consistent with the expectations of metrical phonological theory and expressible through the representational primes and normal computational mechanisms of that theory. Second, the analysis must not only derive the surface position of accent in individual words, but also show how these properties arise through the combination of underlying specifications on morphemes and a system of general rules. In particular, it will not be sufficient simply to assert that a particular stem (= R + S) has a particular pattern of accentual behavior, but rather it must also be shown how the derivational (stem-producing) morphology itself generates the range of accent classes from the underlying properties of the pieces that compose these stems. Finally, the proposed grammar must generate (to some reasonable approximation) only those surface patterns that are exhibited in the data, and exclude unobserved patterns by making them simply ungenerable in the first place.

Before turning to an analysis of the hysterokinetic and proterokinetic types, I will first introduce the formal theory of metrical phonology to be employed.

2.3   Simplified Bracketed Grid Theory

Idsardi's (1992) Simplified Bracketed Grid Theory (SBGT) of metrical phonology provides a useful starting point for the formal analysis of PIE accentuation because it implements certain key properties of the system of Russian accentuation in a direct and appealing way. As in many prior analyses of "irregular" stress, Idsardi encodes the morpheme-specific property of underlying accent through prespecification of metrical structure, in particular through a left-constituent boundary (parenthesis) prespecified to the left of the line 0 grid mark projected by the "accented" vowel. In addition, an Edge Marking Rule places a right parenthesis at the end of line 0. Grids representing the Russian examples in (2) are shown below:

(5)   (x   (x
          x x x   x)   x x x (x)
      a. skóvorod-y   b. skovorod-á̱   'frying pan' GEN.SG/NOM.PL
                                        x

          x   (x   x
      c. (x   x   x)   d. (x   x   (x)
          kómnat-y   kómnat-a̱   'room' GEN.SG/NOM.PL

SBGT crucially defines a metrical constituent ("foot") as an unbroken sequence of grid marks either to the left of a right parenthesis or to the right of a left parenthesis. In (5a) *skóvorody*, the line 0 marks compose a constituent by virtue of the final right parenthesis; no parenthesis matching is necessary. By hypothesis constituents on line 0 are left-headed in Russian, so the insertion of the final right parenthesis generates initial accent in words containing no accented morphemes. In (5b) *skovorodá*, the "accented" property of the suffix is encoded through prespecification of (to the left of the grid mark projected by -*a*. In consequence there is only one constituent in the word, consisting of the final syllable alone; the leftmost grid mark—indeed, the only mark—in this constituent is thus its head, and receives surface accent. In (5d) *kómnata*, two constituents end up being generated on line 0; in such cases the head of the leftmost constituent receives surface stress. Formally this is obtained by construction a left-headed constituent on line 1.

It is easily seen that the PIE examples in (6) are parallel to the Russian ones in (5):

(6)   (x   (x
       x   x   x)   x   x (x)
   a. léi-mon-es   li-mn-ḗi   Amphikinetic: 'lake' NOM.PL/DAT.SG
      (x   (x
      (x   x)   (x   (x)
   b. nókʷt-es   nḗkʷt-ei   Acrostatic: 'night' NOM.PL/DAT.SG

The PIE locative singular null affix does not project a grid mark but still has a prespecified (:

(7) (x (.)   x   x (.)
     nékʷt-Ø 'night' LOC.SG.   li-món-Ø 'lake' LOC.SG

The retraction of accent onto the final syllable in *li-món-Ø* 'lake' LOC.SG has an exact formal parallel in the postaccenting stems in Russian, which, in Idsardi's approach, have a prespecified left parenthesis (to the right of their rightmost grid mark. As shown in (8), if the suffix surfaces as Ø, stress appears on the stem-final syllable in a postaccenting stem such as *karandaʃ-* 'pencil', by virtue of a Retraction Rule (9):

(8) x x   x (x)   x x   x(.)   →   x x (x .)
     karandaʃ-ý   karandaʃ-Ø   karandáʃ
     'pencil' GEN.SG   'pencil' NOM.SG

(9) Retraction: x (.   →   (x .

2.4   Hysterokinetic and Mesostatic Stems

With these formal tools in mind, we can now approach an analysis of the hysterokinetic and mesostatic stems, examples of which are shown in (10):

(10)	*Hysterokinetic*	*Mesostatic*
SG.NOM	*dʰug-h₂tér-s → *dʰugh₂tḗr	*snus-ó-s
ACC	*dʰug-h₂tér-m̥	*snus-ó-m
DAT	*dʰug-h₂tr-éi	*snus-ó-ei
LOC	*dʰug-h₂tér-Ø	*snus-ó-i
PL.NOM/VOC	*dʰug-h₂tér-es	*snus-ó-es

Descriptively, a hysterokinetic stem such as *dʰug-h₂tér-/-h₂tr-* 'daughter' has surface accent on an underlyingly accented desinence (R + S + É), as in DAT.SG *dʰug-h₂tr-éi*. This seems to suggest that a hysterokinetic stem is underlyingly unaccented. Nevertheless, when the desinence is also underlyingly unaccented, as in NOM.PL *dʰug-h₂tér-es*, surface accent does not recede to the initial syllable as one would expect for an unaccented stem like *lei-mon-/*li-mn-* (cf.

6a), but rather appears on the suffix (R + Ś + E). Supposing instead, then, that the suffix here is underlyingly accented, the following metrical structures will be derived on line 0:

(11)   x   (x x)   x (x (x
        dʰug-h₂tér-es   dʰug-h₂tr-éi   'daughter' NOM.PL/DAT.SG

For surface accent to appear on the final syllable in *dʰug-h₂tr-éi*, PIE requires a rule of Stress Clash Resolution (SCR) in (12) which deletes the accent of the first of two consecutive accented morphemes, as shown in (13).[2]

(12) *Stress Clash Resolution*
        (x   (x   →   x   (x

(13)   x   (x (x)   x   x (x)
        dʰug-h₂tr-éi   →   dʰug-h₂tr-éi   'daughter' DAT.SG

However, although SCR successfully derives the hysterokinetic pattern, it introduces a number of problems elsewhere. For example, it would appear to make it impossible to derive even the pattern of a monosyllabic accented stem such as *nókʷt-/*nékʷt-* 'night' where the root's accent would incorrectly be deleted before an accented ending such as DAT.SG *-éi*, giving incorrect **nekʷ-t-éi* (14a). Likewise a mesostatic stem such as *snus-ó-* will also incorrectly yield its underlying accent to that of a following suffix, giving **snus-o-éi* instead of *snus-ó-ei* (14b):

(14)   (x   (x)   x   (x)
        a. nekʷ-t-ei   →   **nekʷ-t-éi
             x (x(x)   x   x(x)
        b. snus-o-ei   →   **snus-o-éi

In fact, unless *snus-ó-* and *dʰug-h₂tér-/-h₂tr-* have a distinct metrical pre-specification—compare (14b) and (13)—they cannot possibly have distinct metrical behaviors. The question then amounts to which of them has a different or more complex structure. What I will suggest here is that the anomaly lies not, in fact, with the hysterokinetic type, but rather with the acrostatic and mesostatic types, which *fail* to undergo SCR (12). This is the first respect in which the solution offered here departs significantly from previous approaches, which have taken the fixed accent types to be as unexceptional in PIE as they are in, say, Russian or Lithuanian. We will see, however, that this reorientation of perspective is not only perfectly consistent with the essential tenor of Idsardi's analysis of Russian, but in addition affords the possibility of extending it to the full range of complex patterns observed in PIE.

Specifically, I propose that instead of having a prespecified left parenthesis (, acrostatic stems in fact are prespecified with a right parenthesis) to the right of their initial, accented syllable. Thus the first syllable of the word must end a metrical constituent; this constituent will always be first in the word, and so surface accent will always be initial. The difference between acrostatic *nék^wt-ei* 'night' DAT.SG (15a) and hysterokinetic *d^hug-h₂tr-éi* 'daughter' DAT.SG (15b) is illustrated below:

(15)   (x

       x) (x)
    a. nék^wt-ei

                             (x

     x (x (x)   x x (x)
    b. d^hug-h₂tr-éi → d^hug-h₂tr-éi

Crucially the representation in (15a) will now not meet the structural description of SCR (12), since the rule makes reference to two left parentheses, and only one is present.

Whereas a morpheme such as DAT.SG *-ei*, whose left edge is prespecified with a left parenthesis, places accent at its left edge, a morpheme such as *nók^wt-/*nék^wt-*, whose right edge is prespecified with a right parenthesis, instead positions accent *as far to the left as possible* (assuming, of course, left-headed constituents on line 1). To distinguish such morphemes from ordinary "accented" morphemes, here I will refer to this type of morpheme as *back-accenting.*

To derive the mesostatic *snus-ó-*, it is not sufficient to make the suffix simply back-accenting, since this will incorrectly force accent to the initial syllable as shown in (16a). Instead a type of prespecification is required that is not only immune to SCR (12), like the back-accenting (acrostatic) type, but that also fixes stress at a certain noninitial location, like an ordinary accented desinence. This behavior can be derived provided that the affected syllable is prespecified with both) and (on line 0, as shown in (16b):

(16)   (x (x   (x x

        x x)   x x)x)   x x) (x)
    a. snus-o-   snús-o-es   **snús-o-ei   Incorrect
                   (x x   (x x

        x (x)   x (x)(x)   x(x)(x)
    b. snus-o-   snus-ó-ei   snus-ó-ei   Correct

This medial fixed accent simply results from the simultaneous presence of two independently necessary prespecifications: (forward-)accenting, by virtue of a left parenthesis (, and back-accenting by virtue of a right parenthesis).[3] Note again that **snus-ó-ei*, like **nékʷt-ei*, does not meet the structural description of SCR (12) owing to the right parenthesis in its prespecified structure.

Grids representing the four accent types analyzed so far are now shown below, first with an unaccented suffix (NOM.PL *-es*) in (17a) and then with an accented suffix (DAT.SG **-ei*) in (17b):

(17)   *Acrostatic*   *Amphikinetic*   *Mesostatic*   *Hysterokinetic*

	x) x)	x x x)	x (x)x	x (x x)
a.	nókʷt-es	léi-mon-es	snus-ó-es	dʰug-h₂tér-es
	x) (x)	x x(x)	x (x) (x)	x [x (x)
b.	nékʷt-ei	li-mn-éi	snus-ó-ei	dʰug-h₂tr-éi

It is easily verified that only the grid for **dʰug-h₂tr-éi* meets the structural description of SCR (12) since in no other instance does line 0 contain precisely (x (x. The left parenthesis deleted by SCR in **dʰug-h₂tr-éi* is indicated with a square bracket in (17b).

2.5 Proterokinetic Stems

I turn now to the proterokinetic stems, which present a greater analytic challenge. Three examples are provided below:

(18)

	'widow'	'thin' MASC	'thin' FEM
SG.NOM	**h₁u̯idʰ-éu-h₂-Ø*	**ténh₂-u-s*	**tn̥h₂-éu̯-ih₂-Ø*
DAT	**h₁u̯idʰ-u-éh₂-ei*	**tn̥h₂-éu̯-ei*	**tn̥h₂-u-i̯éh₂-ei*
PL.NOM/VOC	**h₁u̯idʰ-éu-h₂-es*	**ténh₂-eu-es*	**tn̥h₂-éu̯-ih₂-es*
LOC	**h₁u̯idʰ-u-éh₂-su*	**tn̥h₂-ú-su*	**tn̥h₂-u-i̯éh₂-su*

Surface accent falls on the syllable before an underlyingly accented ending, as in **tn̥h₂-u-i̯éh₂-ei* 'thin' FEM DAT.SG, but on the syllable *before* the syllable before an underlyingly unaccented ending, as in **tn̥h₂-éu̯-ih₂-es* 'thin' FEM-NOM.PL. It is important to notice that in the latter case, surface accent does not in fact recede to the initial syllable, as in Russian *skóvorod-y* (5a), but instead appears on the syllable nucleus of the third-to-last morpheme.

The proterokinetic pattern begins to make sense once it is compared with the hysterokinetic pattern. Specifically, the alternation of accent position in the proterokinetic type is identical to that of the hysterokinetic type, except that the surface accent is *one syllable further to the left:*

(19)

	'thin' MASC	'daughter'
SG.NOM	**ténh₂-u-s*	**dʰug-h₂tér-s* (→ **dʰugh₂tḗr*)
DAT	**tn̥h₂-éu̯-ei*	**dʰug-h₂tr-éi*
PL.NOM/VOC	**ténh₂-eu-es*	**dʰug-h₂tér-es*
LOC	**tn̥h₂-ú-su*	**dʰug-h₂tr̥-sú*

It is also significant that in the proterokinetic stems stress *never falls on the final syllable,* and this suggests that the final syllable is metrically invisible. The analysis I present will capitalize on both these observations.

Phenomena in which peripheral elements are immune to or excluded from phonological computations ("extrametricality") are of course quite common. Beyond this, however, Halle (1998) noted words in English in which stress appears on the antepenultimate syllable even though the penult is heavy:

(20) (x) x .   x (x) x   (x) x .   x(x) x
       Lómbardy vs. Lombárdi   Wáshington vs. malínger

Halle proposed that these words have suffixes (such as *-y*, *-ton*) that are metrically invisible: they simply do not project any line 0 grid mark. Crucially, this metrical invisibility is not the same as ordinary extrametricality, as seen in *Lombárdi* or *malínger*; the penult in *Lómbardy* and *Wáshington* is also external to any metrical constituent ("extrametrical" in a different sense), for if it were not, it would have attracted stress, as in *bambóo,* for example.

For present purposes the importance of Halle's analysis is that it draws a distinction between metrically inert elements that are invisible to the grid, and therefore cannot figure in metrical calculations, versus those that are visible, but, being peripheral in a certain domain, are still excluded from metrical constituents.

With this distinction in mind we can now begin to make sense of the proterokinetic pattern. Specifically, I propose that proterokinetic stems are derived by cyclic, accented suffixes that have the property of preventing the final syllable in the domain from projecting to the grid. Formally this is obtained by having these suffixes associated with a rule of Grid Mark Deletion:

(21) *Grid Mark Deletion*
      x] → .]

It will also be necessary to treat Retraction as a cyclic (as well as a noncyclic) rule.

The mechanics of this proposal are best illustrated through derivations. Consider first what occurs on the cycle introduced by a proterokinetic suffix such as *-(e)u-* or *-i(e)h_2-*.[4]

(22) a. [tenh₂-eu]-   b. [[tenh₂-eu]-ieh₂]-

         x (x   x (x   Cycle on *-eu-*
         x (.   x (.   Grid Mark Deletion (21)
        (x .   (x .   Retraction (9)
                        x x (x   Cycle on *-i(e)h₂-* (Stress Erasure)
                        x x (.   Grid Mark Deletion (21)
                        x (x .   Retraction (9)

The derivations for (22a) and (22b) are identical on the first cycle. Grid Mark Deletion occurs, triggered by the proterokinetic suffix *-(e)u-, followed by cyclic Retraction. The output is a stem accented one syllable before *-(e)u-; that is, on the Root. In (22b) a second cycle on *-i(e)h$_2$- occurs: Stress Erasure removes metrical structure created on the first cycle and the same steps are repeated to give a stem accented on the syllable before *-i(e)h$_2$-, namely *-(e)u-.

The stems created by the cyclic phonology are then inputs to the noncyclic phonology in the domain created by the desinences. The derivations in (23) show the behavior of *t(e)nh$_2$-(e)u- in the context of an unaccented suffix (NOM.PL *-es) and an accented suffix (DAT.SG *-ei):

(23) a. tenh₂-eu-es   b. tenh₂-eu-ei

(x x x	(x x (x		Noncyclic
(x x x)	(x x (x)		End Rule Right
(x x .)	(x x (.)		Grid Mark Deletion (21)
n/a	(x (x .)		Retraction (9)
n/a	x (x .)		SCR (12)
*ténh₂-eu-es	*tn̥h₂-éu-ei		

At the beginning of the noncyclic phonology, projection of grid marks occurs again, at which point the suffix *-(e)u- reprojects a grid mark, and, since this mark is no longer final in its domain, it is not subject to Grid Mark Deletion; instead the final syllable's grid mark is deleted. In (23b) Retraction applies; this then feeds SCR, which in turn deaccentuates the Root, leaving surface accent on *-(e)u-.[5]

For descriptive convenience I will call these proterokinetic suffixes *extrametricalizing* and notate them with the diacritic [EM]. A summary of the underlying properties of the metrical grids of stems five accent types is shown in (24).[6] (The reader may also find it useful to consult the additional list in (34) and (35) at the end of the chapter:)

(24)

		x)
a. Acrostatic	nók^{w}t-	
		x x
b. Amphikinetic	lei-mon-	
		x (x)
c. Mesostatic	snus-o-	
		x (x
d. Hysterokinetic	d^{h}ug-h₂ter-	
		x (x
e. Proterokinetic	tenh₂-eu[EM]-	

2.6 Putting the Pieces Together: Stem Derivation and Accent Class

Comparatively little attention has been given to how the accent class of a stem results from the contribution of the underlyingly specified properties of its parts, but some observations commonly encountered in the literature are given here:

(25) a. Amphikinetic stems: *-ios- elative adjectives ("unusually X"); zero-derived neuter collectives of stems in *-m(e)n-

    b. Hysterokinetic stems: nouns with Suffix *-$h_2t(\acute{e})r$-; active participles in *-$(\acute{o})nt$-

    c. Proterokinetic stems: nominalizations in *-$t(\acute{e})u$- and *-$t(\acute{e})i$-; neuter nouns in *-$(\acute{e})n/r$-, *-$(\acute{e})u$-, and *-$(\acute{e})i$-; adjectives in *-$(\acute{e})u$-; derived feminine adjectives in *$\underset{\,}{i}(\acute{e})h_2$-

    d. Mesostatic stems: verbal adjectives in *-$t\acute{o}$-, *-$n\acute{o}$-, *-$w\acute{o}$-; imperfectives in *-$sk\acute{e}$-

In all the cases in (25) it is the suffix that determines the accent class of the stem; the root, on the other hand, appears to make no contribution aside from selecting the specific affix allomorph. This fact emerges as a natural consequence of the assumption of the analysis presented here. Consider (26):

(26)

Suffix	Unaccented root or cyclic suffix	Accented root and noncyclic suffix
a. x	Amphikinetic	Acrostatic
b. (x	Hysterokinetic	Acrostatic
c. $(x_{[EM]}$	Proterokinetic[7]	Acrostatic
d. (x)	Mesostatic	Acrostatic

The accent class that will be derived for a stem (R+S) depends on three properties: (i) the accentuation (i.e., metrical prespecification) of the Root; (ii) the metrical prespecification of the Suffix; and (iii) whether the Suffix is cyclic or noncyclic. If the Suffix is noncyclic, it will not trigger Stress Erasure and so will never be able to override the accent of an accented Root; thus the right column in (26) shows that an accented Root and noncyclic Suffix will always give an acrostatic stem. More interesting for our purposes is the left-hand column, which shows the result when either the Root has no underlying accent (prespecified structure) *or* if the Suffix is cyclic, in which case Stress Erasure will obliterate any metrical prespecification present on the Root.

For example, amphikinetic collective/abstract neuter nouns, such as *$s\acute{e}h_1$-mon/men- 'seed stuff' (27), can be derived by zero affixation to a proterokinetic neuter stem denoting an individual, such as *$s\acute{e}h_1$-$m\underset{\,}{n}$/*sh_2-$m\acute{e}n$- '(a) seed' (28). On the present analysis the change in accent class arises because the collective/abstract suffix -Ø is both cyclic and unaccented. Assuming that the End Rule Right and line 1 constituent construction are *not* cyclic rules, a cyclic unaccented suffix will automatically create an unaccented stem:[8]

(27) séh₁-mn$_{[EM]}$-Ø   seh₁-men$_{[EM]}$-ei →   seh₂-mén-ei
     (x   x   .)   (x   x   (.))   x   (x .)
     *séh₁-men   *sh₂-mén-ei
     '(a) seed' NOM.SG   '(a) seed' DAT.SG

(28) [[séh₁-men$_{[EM]}$]-Ø]-Ø]   [[seh₁-men$_{[EM]}$]-Ø]-ei]
     (x   .   (x   .   Cycle on -*men*-
       x   x   .   x   x   .   Cycle on -Ø$_{coll}$
       x   x   .   .)   x   x   .   (x)   Noncyclic
     *séh₁-mon → *séh₁mō   *sh₁-mn-éi
     'seed stuff' NOM.SG   'seed stuff' DAT.SG

2.7 Additional Derived Accent Types

In addition to suffixes that create derived stems belonging to the accent classes in (26), there are other suffixes in PIE that create acrostatic stems (with fixed initial accent), crucially irrespective of the accentuation of the base. One such derivation occurs for many verbal roots with imperfective Aktionsart that have a perfective stem with the suffix *-s-*, traditionally called the "sigmatic aorist" stem. These derived perfective stems invariably have fixed root stress with lengthened grade allomorph (*ē*) vowel alternating with (*e*). For example, the imperfective stem *kél-* ~ *kl̥-* 'be driving on' forms a derived perfective *kḗl-s-* ~ *kél-s-* 'drove on'. Whereas *kél-* ~ *kl̥-* is amphikinetic (29a), with surface accent alternating between the initial syllable (with an unaccented singular desinence such as *-mi)* and the final syllable (with an accented plural desinence such as *-te*), *kḗl-s-* ~ *kél-s-* is acrostatic, with fixed initial stress (29b):

(29) a. *kél-mi* 'I am driving on'; *kl̥-té*   'you (PL) are driving on'
     b. *kḗl-s-m̥* 'I drove on';   *kḗl-s-te* 'you (PL) drove on'

The introduction of the back-accenting (right-parenthesis) prespecification for acrostatic stems in section 2.4 now makes possible a completely straightforward analysis of this type: the suffixes that create derived acrostatic stems are both *back-accenting* and *cyclic*. As cyclic affixes, they will erase any accentual properties of the base. As back-accenting, they force accent to the initial syllable just as in acrostatic stems.

(30) x   x x)
    kel-   'be driving on'   kḗl-s-   'drove on'
  (x   (x   x
   x   x)   x x) x)
    kél-mi   'I drive on'   kḗl-s-m̥   'I drove on'
   (x   (x   x
   x (x)   x x)(x)
    kl̥-té   'you (PL) drive'   kḗl-s-te   'you (PL) drove on'

In addition, we now predict that there should be another possible suffix type:
back-accenting and *noncyclic*. These suffixes should have a slightly different
behavior. Instead of erasing the accent of the base and moving accent to the
initial syllable, noncyclic back-accenting morphemes move accent to the initial
syllable only with unaccented bases, but leave accent where it is on accented
bases.

It turns out that this behavior is exactly that of the subjunctive suffix **-elo-*.
The indicative and subjunctive forms of the unaccented verb root stems **h₁és-/*
**h₁s-* 'is' and **gʷém-/*gʷm-* 'took a step' are illustrated in (31). In the left
column, in the indicative, the surface accent alternates, depending on whether
the agreement desinence is underlyingly unaccented (3SG) or accented (2PL),
showing that the root stems are unaccented. In the right column, in the subjunctive, accent is uniformly on the initially syllable.

(31)   *Indicative*   *Subjunctive*
  3SG   *h₁és-ti   *h₁és-e-ti   'is'
  2PL   *h₁s-té   *h₁és-e-te   'you (PL) are'
  3SG   *gʷém-t → *gʷemd   *gʷém-e-ti   'she/he took a step'
  2PL   *gʷm̥-té   *gʷém-e-te   'you (PL) took a step'

In contrast, in (32) the stem is a derived imperfective **gʷm̥-sḱé-* formed with
the cyclic accented imperfective suffix **-sḱé-*, which makes a mesostatic stem
with accent fixed on **-sḱé-* irrespective of the accent of the desinence. In the
subjunctive forms on the right, the position of the accent does not change as
it it did in (31); rather, the accent remains fixed on **-sḱé-*.

(32)   *Indicative*   *Subjunctive*
  3SG   *gʷm̥-sḱé-ti   *gʷm̥-sḱé-e-ti   'she/he is stepping/going'
  2PL   *gʷm̥-sḱé-te   *gʷm̥-sḱé-e-te   'you (PL) are stepping/going'

Put more briefly, subjunctive **-elo-* causes unaccented stems to become acrostatic (33a–b), but accented stems are unchanged (33c–d):

(33)   (x   x   x   (x x
              x   x) x)   x   (x)x) x
           h₁és-e-ti   gʷm̥-sk̂é-e-ti
        a. 'is' SUBJ   c. 'is stepping' SUBJ

             (x   x   (x x   x
              x   x)(x)   x   (x)x)(x
           h₁és-e-te   gʷm̥-sk̂é-e-te
        b. 'you (PL) are' SUBJ   d. 'you (PL) are stepping' SUBJ

Because *-e/o- is noncyclic it does not affect the accent of the base to which it attaches. As shown in (33c–d), if the base is accented, it remains so. If the base is unaccented, however, the right parenthesis prespecified on *-e/o- creates the right edge of a metrical constituent whose leftmost mark is the initial syllable, which then carries surface accent. Crucially, this holds even when the desinence is underlyingly accented, as in (33b).

2.8 Conclusion

Underlyingly accented and unaccented morphemes (syllables) are known from PIE descendants that retain traces of the archaic system, such as Russian, Lithuanian, and Vedic Sanskrit; following Idsardi 1992 it is normally assumed that in these languages an accented morpheme is prespecified with a left parenthesis in its metrical grid. Somewhat more complex patterns are observed in PIE, however, and to derive them I have proposed two additional prespecification types: back-accenting, which contains a right parenthesis, and accented extrametricalizing, which has a left parenthesis and moreover deletes the final grid mark in its domain. The traditional "fixed" accent types such as the acrostatic and mesostatic patterns were shown to result from a combination of "accented" and "back-accenting" prespecification; the hysterokinetic type results from a general rule of Stress Clash Resolution and is proposed to be relatively unmarked in PIE. Finally, cyclic and noncyclic modes of derivation not only ensure the proper surface distribution of accent in morphologically complex stems but also *explain* how (and why) the accent class of a stem is determined by properties of its root morpheme and its suffix or suffixes.

To summarize, a list of the different root and suffix types with examples of each is given in (34) and (35):

(34) Cyclic
    a. x   Unaccented   Abstract/collective neuter *-Ø:
                                          amphikinetic
    b. (x   Accented   Participial *-ónt-/*-nt-: hysterokinetic
    c. x)   Back-accenting   Perfective *-s-: derived acrostatic
    d. (x)   Fixed accenting   Imperfective *-skḗ-: derived mesostatic
    e. (x[EM]   Extrametricalizing Proterokinetic stem suffixes (e.g., *-(é)u-)

(35) Noncyclic
    a. x   Unaccented   Direct case desinences, singular verb
                                            agreement; unaccented roots
    b. (x   Accented   Oblique case desinences, plural verb
                                            agreement
    c. x)   Back-accenting   Subjunctive *-e/o-; accented roots[9]
    d. (x)   Fixed accenting   Simple mesostatic stems with theme
                                            *-ó- (e.g., snusós)

Morphemes as in (34) and (35) combine to correctly produce stems with all
the standard PIE accent classes. But a more important result—and a singular
advantage of the present proposal—is that these accent classes are the only
ones that can be generated from these underlying forms. From this it becomes
clear that explicit theoretical analysis does not merely respond to historical
data obtained through conventional descriptive techniques, but is in fact essential for understanding the nature of reconstructed grammatical systems.

Notes

1. As an anonymous reviewer points out, this holds, of course, only if reconstructed PIE
does indeed represent the grammar of some idealized speaker at some specific point in
time (i.e., to the extent that the different accent types under discussion can reasonably be
assumed to have been synchronically simultaneous), a nontrivial assumption.

2. Stress Clash Resolution processes are, of course, quite common crosslinguistically,
but more specifically, a rule similar to (12) in SBGT can be found, for example, in
Halle's treatment of stress deletion under clash in words such *eleméntary* or *compúlsory*
in American English (Halle 1998, 558). Although secondary stress normally appears
on the suffixes *-ary* and *-ory* (cf. *légendàry, ínventòry*), there is no secondary stress in
posttonic position.

3. It is clear that acrostatic stems such as **nókʷt-/*nékʷt-* could also be underlyingly
specified with both a right and a left parenthesis, although a single right parenthesis is
sufficient to derive the initial accent position.

4. I assume that the root is a noncyclic domain; in other words, stress rules do not
apply to the root domain on a zeroth cycle. As Halle and Kiparsky (1977) show for
Vedic Sanskrit, a stem whose last cyclic suffix is unaccented will surface with fixed

initial accent. For Halle and Vergnaud (1987) this occurs because, effectively, the End Rule Right (followed by left-headed line 1 constituent construction) applies cyclically, creating a representation like that of *léimones* in (6a); any subsequent noncyclic suffixes will then be powerless to move the accent off the stem-initial syllable. On these assumptions, if roots were cyclic domains, an unaccented root with no cyclic suffix would produce an acrostatic stem with fixed initial accent—clearly an incorrect result. Nevertheless, it is noteworthy that the Vedic pattern does not in principle require all metrical rules to be cyclic: it is only End Rule Right (and line 1 constituent construction) that must apply cyclically. As discussed in section 2.6, however, the present proposal is that in PIE the End Rule Right applies only noncyclically. As a result, there is no compelling evidence either for or against the hypothesis that roots are cyclic stress domains in PIE.

5. An anonymous reviewer raises the important question of what defines the domain for Grid Mark Deletion. For example, if a proterokinetic stem were to have two (or more) noncyclic affixes, then the rule as stated would appear to incorrectly delete the grid mark of the *last* one, instead of the first one. I am not aware of any data bearing on this question, so for now I leave the issue open.

6. If, in principle, any type of suffix could carry the extrametricalizing diacritic, then two additional types of extrametricalizing suffixes are predicted to occur: (i) noncyclic and accented, which produces a surface pattern identical to the mesostatic type; and (ii) unaccented, which, when noncyclic and attaching to an unaccented stem, or when cyclic (irrespective of the accent of the stem), will give a pattern of initial accent with an unaccented desinence but penultimate accent with an accented desinence. This second pattern is not attested to my knowledge (but would only be distinguishable from a normal proterokinetic pattern for stems of three syllables or more); thus we will assume that the extrametricalizing type is always accented, for reasons yet to be determined.

7. Or mesostatic, if simultaneously the suffix is noncyclic and the root is unaccented: see note 6.

8. I assume that certain metrical rules apply cyclically, but that crucially the End Rule Right does not, or else unaccented cyclic suffixes would produce stems with fixed initial accent, as in Vedic (Halle and Kiparsky 1977), and sh_1-*mn-éi* 'seed stuff' DAT.SG would incorrectly surface with initial accent. The grammar proposed here for PIE apparently represents a middle ground between the Vedic system (all metrical rules are both cyclic and noncyclic) and the Lithuanian one (metrical rules are noncyclic only), as analyzed by Halle and Vergnaud (1987).

9. Accented roots may also be of type (35d); see note 3.

3 Agree and Fission in Georgian Plurals

Martha McGinnis

3.1 Introduction

One of Morris Halle's many lasting contributions to the study of morphology is his elaboration of a principled relationship between morphology and syntax. In particular, his work in Distributed Morphology argues that syntactic nodes provide the domains within which morphological disjunctivity obtains (Halle and Marantz 1993, 1994; Halle 1997a, inter alia). More specified Vocabulary Items outrank less specified ones in the competition to spell out or *discharge* the features of a given syntactic node. In the usual case, only one item is inserted into a given node; in this case, all items competing for insertion are in a disjunctive relationship. In some cases, however, the node may undergo *fission*, allowing the additional insertion of a lower-ranked item to discharge features not discharged by the first item inserted. The operation of fission relaxes the strict disjunctivity among items competing for insertion into the same node; it also allows disjunctive blocking to apply across traditional "position classes." Noyer (1992) refers to this phenomenon as *discontinuous bleeding*.

Béjar (2003) points out that disjunctive blocking can also arise from the syntactic operation *Agree*. In the simplest case, an agreeing head probes the structurally closest matching constituent, blocking more distant constituents from triggering agreement. English subject agreement works like this, always probing the (underlined) highest argument in the clause:

(1) a. <u>I</u> am/*is the boss.

    b. <u>The boss</u> is/*am me.

Béjar argues that agreement can be generalized (e.g., person, number), or specified (e.g., first person, plural number). She postulates that generalized agreement probes the closest matching constituent, as in (1), while specified agreement has a broader reach: if the closest argument does not match the specified feature, the head can probe a second time.[1] In one illustration of the theory, Béjar argues that the person-marking prefixes on Georgian verbs arise from specified [π [Participant]] agreement; see McGinnis 2008 for additional arguments to this effect.

In this chapter, I demonstrate that an elusive interaction in Georgian inflection can be elegantly captured under a Distributed Morphology analysis that combines Halle's (1997a) theory of fission with Béjar's theory of Agree. The interaction in question is the disjunctivity among number-agreement suffixes on Georgian verbs. A key instance of this interaction arises between plural *-t* and the third-person plural Tense/Aspect/Mood (TAM) suffix, traditionally called the "screeve marker." The two suffixes occupy different positions, as illustrated by (2), where the plural *-t* follows the TAM suffix, here *-a*.

(2) g-nax-a-t
     2.DAT-see-AOR-PL
     'He/she saw you (PL).'

Nevertheless, double plural marking is impossible, even when the two plural suffixes would be discharging different plural features, as in (3). The suffix *-es* indicates a third-person plural subject (3a), but *-t* cannot discharge the plural feature of a second-person object in (3b), even though it does so in (2). I argue that this interaction arises from the nature of specified Agree. Following Béjar 2003, I postulate that a Georgian clause has only one number-agreement feature, which is specified as plural; thus, only a single plural argument (here, the subject) can trigger number agreement on the verb.

(3) a.   g-nax-es
          2.DAT-see-AOR.3PL
          'They saw you (SG/PL).'
     b.   *g-nax-es-t
          2.DAT-see-AOR.3PL-PL
          'They saw you (PL).'

The plural *-t* in (3b) also cannot show agreement with a plural TAM suffix, if both suffixes in (3b) are taken to show agreement with the plural subject (on the reading 'They saw you (SG)'). I argue that this interaction arises because the so-called TAM suffix and the plural *-t* actually both discharge the features of the same node, via fission. If the plural feature is not discharged by the TAM suffix, it fissions off from the TAM head and is discharged by *-t*; if it is discharged by the TAM suffix, *-t* cannot be inserted.

In the next section I sketch an explicit theory of morphosyntactic features based on Harley and Ritter 2002, and outline the main analysis in more detail.

Section 3.2.1 lays out the role of syntactic number agreement on T specified for the plural feature [Group], as well as its interaction with clitic movement of the first- and second-person pronouns. As I will show, number agreement in Georgian is also affected by the application of an Impoverishment operation to the [Group] feature of first-person dative plural pronouns. Vocabulary items for the person-marking prefixes are provided as well. Section 3.2.2 describes fission of the TAM node and provides vocabulary items for an illustrative subset of the number-marking suffixes. Evidence for the proposed analysis is provided in section 3.3. Section 3.3.1 argues that only one argument per clause triggers number agreement, even though number marking may be specified in either of two suffix positions. I demonstrate that the analysis outlined in section 3.2 provides a straightforward account of a formerly mysterious inter-action between plural *-t* and a third-person suffix *-s*, also discussed by Ander-son (1984). Section 3.3.2 discusses an empirical challenge to the proposed analysis—namely, clauses in which both a dative first-person plural argument and a nominative plural argument appear to trigger number agreement. Drawing on crosslinguistic evidence, I argue that dative first-person plurals in Georgian are morphologically distinguished from their singular counterparts only by a special collective first-person feature [Multispeaker], not by [Group], and that therefore there are actually no forms showing [Group] agreement for more than one argument.

3.2   The Analysis: Agree and Fission

Like Béjar (2003), I assume a slightly modified version of Harley and Ritter's (2002) geometry of privative morphosyntactic features. The relevant features are illustrated in (4). Nothing in the analysis depends on assuming privative features: I simply adopt this as the more restrictive hypothesis. Note that Harbour (chapter 8, this volume) argues in favor of a binary feature system instead.

(4)

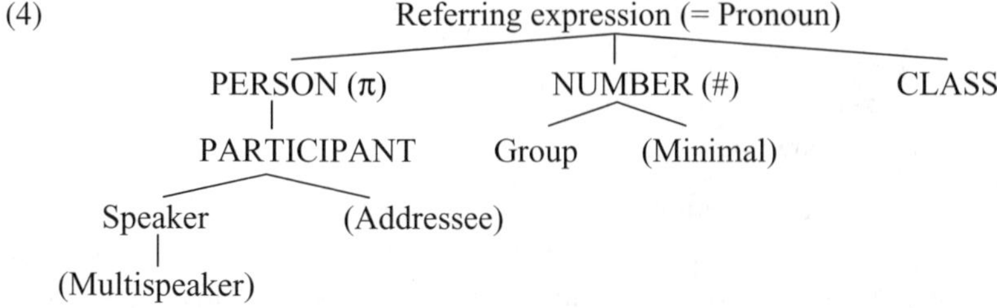

The major organizing nodes are [Person], [Number], and [Class]. I have not included details of the [Class] node, but it is associated with animacy (sen-tience) and grammatical gender. First and second person both have the feature

[Participant]. Crosslinguistically, [Speaker] is used for first person and [Addressee] for second person; similarly, [Minimal] is used for singular and [Group] for plural. However, I assume that the features [Addressee] and [Minimal] are activated only in systems that have an inclusive/exclusive first-person distinction, and a dual/plural number distinction, respectively. Since this is not the case in Georgian, I assume that neither [Addressee] nor [Minimal] is activated in this language; instead, second person and singular are represented by underspecified [Participant] and [#] nodes, which are interpreted by contrast with their more specified [Speaker] and [Group] counterparts (McGinnis 2005). Third person is represented by an unspecified [π] node. In addition to the features proposed by Harley and Ritter, I also postulate a special "multiple-speaker" feature [Multispeaker], a dependent of [Speaker], which can express first-person plural in some languages. See section 3.3.2 for more discussion.

3.2.1   Agree: Specified Number Agreement on T

Following Béjar 2003, I postulate that number agreement on Georgian verbs derives from a single uninterpretable [# [Group]] feature, specified on T. T begins by probing the closest argument. If this argument (the subject) is plural, T agrees with it, blocking number agreement with lower arguments, as seen in (3), repeated here as (5a). If not, T probes a second time. In that case, if the object is plural, T can agree with it, as seen in (2), repeated here as (5b). Otherwise, the [Group] feature deletes from the [#] node, and no plural marking appears (5c).[2]

(5) a. g-nax-es(*-t)
         2.DAT-see-AOR.3PL
         'They saw you (SG/PL).'
     b. g-nax-a-t
         2.DAT-see-AOR-PL
         'She/he saw you (PL).'
     c. g-nax-a
         2.DAT-see-AOR
         'She/he saw you (SG).'

An additional clarification is in order. Béjar argues that a feature on a head H can probe downward only once—namely, when H merges at the root of the derivation. After that, as the tree extends, H probes upward.[3] This characteristic suggests a possible account of an additional restriction on number marking—namely that, unlike participant (first- and second-person) objects, third-person objects cannot trigger plural agreement, even when the subject is singular. (6a) shows that the third-person plural object cannot trigger the third-person plural TAM suffix shown in (6a); it also cannot trigger the default plural *-t* (6b). Only an ambiguous form is possible, as in (6c).

(6) a. *v-nax-es
     1-see-AOR.3PL
     'I saw them.'

   b. v-nax-e-t
     1-see-AOR.PART-PL
     *'I saw them.' *(Fine on the readings 'We saw him/her/it/them.')*

   c. v-nax-e
     1-see-AOR.PART
     'I saw him/her/it/them.'

This generalization holds even in clauses with a dative subject—either subject-experiencer clauses, or those in perfect or pluperfect aspect. In clauses with a nominative or ergative subject, the TAM suffix always agrees with the subject, showing person agreement and third-person number agreement. On the other hand, in clauses with a dative subject, the TAM suffix agrees in person with the nominative object, but neither the TAM suffix nor the plural *-t* shows number agreement with a third-person nominative object (Nash 1994, 169).

This distinction between third-person and participant objects arises for syntactic reasons. Halle and Marantz (1993) argue that first- and second-person arguments in Georgian are syntactic clitics, which move out of their base positions, while third-person arguments remain in situ.[4] I propose that these clitics attach to T, allowing the [Group] feature on T to target them, as in (5b). By contrast, a third-person object will remain below T, inaccessible for agreement with it because T cannot probe downward a second time—hence the ill-formedness of (6a).[5] Default (here, phonologically null) number agreement appears instead, as in (6c). I assume that the syntactic clitics are phonologically null and trigger agreement on *v* and T, following Béjar's (2003) analysis (see McGinnis 2008 for further evidence). If the clitics must move through spec-*v*P on the way to T, I assume that they tuck in below the external argument, so that it remains closer to T.[6]

The tree in (7) shows the key features of the proposed syntactic analysis. The T head has both person- and number-agreement features, with the number feature specified with [Group]. The person-agreement feature on T (which manifests itself as person agreement on the TAM suffix) targets the closest nominative or ergative argument, here the external argument, shown in spec-*v*. The number-agreement feature first targets the closest argument, here the external argument; if this argument is plural, it checks the number-agreement feature on T. Otherwise, the number-agreement feature can probe a second time, this time upward to a first- or second-person object clitic, which for the sake of concreteness I have shown undergoing clitic movement to spec-T.

(7)

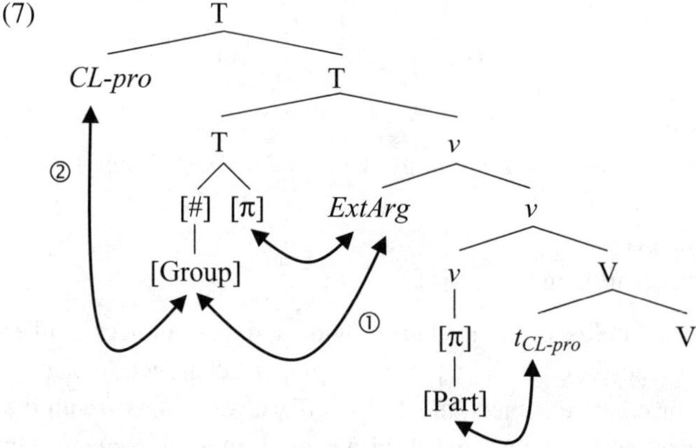

If both the subject and object are first- and second-person clitics, both will move to specifiers of T, assuming that multiple specifiers are possible (Chomsky 1995). It is plausible that third-person external arguments also move to spec-T, although nothing in the proposed analysis hinges on this. I assume underlying SOV order for Georgian, following Nash 1995.

The tree in (7) also shows person agreement on *v*, which is specified for the feature [Participant]. Like the [Group] specification on T, this feature probes downward once, then upward if there are no Participant internal arguments. In other words, it agrees preferentially with a first- or second-person internal argument, otherwise with a first- or second-person external argument. If there are no first- or second-person arguments, the [Participant] feature deletes. According to Béjar 2003, the [π] feature then agrees with a third-person subject; this agreement is spelled out by a null default Vocabulary Item. See Béjar 2003 and McGinnis 2008 for more extensive discussion of person agreement in Georgian verbs.

In section 3.3.2, I argue that plurality in the Georgian dative first person is expressed via person features, rather than number features (Ritter 1997; Harley and Ritter 2002). Specifically, I propose that first-person plural in Georgian has the person feature [Multispeaker], which necessarily co-occurs with the number feature [Group]; but that, in the dative case, the [Group] specification of first-person pronouns undergoes morphological deletion or Impoverishment (Bonet 1991), as in (8). Thus, [Multispeaker] alone expresses number in dative first person.[7]

(8)

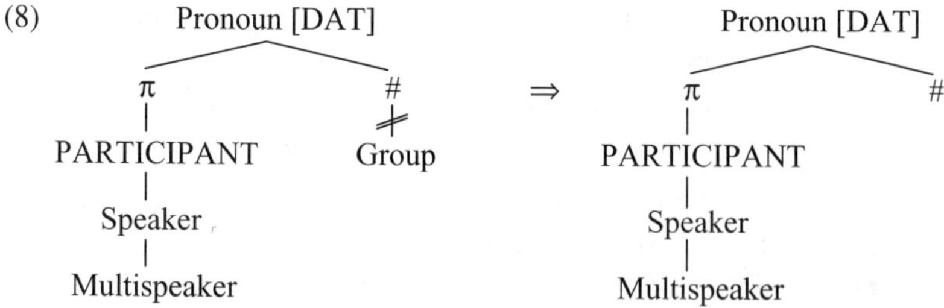

Assuming that the syntactic derivation proceeds phase by phase (Chomsky 2000, 2001), I propose that this Impoverishment operation occurs at the spell-out of the *v*P phase—that is, after dative case is valued, but before number agreement with T.[8] Even though the [Group] node of dative first-person arguments is deleted by Impoverishment, the argument itself remains accessible to syntactic movement, such as the clitic movement in (7). After Impoverishment, it lacks a [Group] feature to agree with the uninterpretable [Group] feature on T, but its [π] features can still value the uninterpretable [π] feature on T.

The prefixal Vocabulary Items for the person agreement on *v* are listed in (9).[9] I assume that a dependent feature is inherently more specified (and thus more highly ranked) than the feature it depends on—for example, (9a) is ranked above (9b) because [Multispeaker] is a dependent of [Speaker]. (9c) and (9d) are not ranked by Pāṇinian disjunctivity, also known as the Elsewhere Condition (Kiparsky 1973; Anderson 1984), but in the proposed analysis, the choice between them is decided by syntactic agreement, since *v* agrees with only one argument. (9c) is inserted if *v* agrees with a second-person dative argument, and (9d) if *v* agrees with a first-person nominative argument. Since this agreement node does not undergo fission, only one of the items in (9) will be inserted into it.

(9) a. [Multispkr, DAT] ↔ /gv-/   1PL.DAT
     b. [Spkr, DAT]   ↔ /m-/   1SG.DAT
     c. [Part, DAT]   ↔ /g-/   2.DAT
     d. [Spkr]   ↔ /v-/   1SG.NOM
     e. [Part]   ↔ Ø (or *x-* in env. *ar* 'be')   2.NOM
     f. elsewhere   ↔ Ø   3

3.2.2 Fission: Interactions among Number-Marking Suffixes

The TAM (screeve) suffix on a Georgian verb reflects tense (aorist, present, future), aspect (imperfect, perfect, pluperfect), mood (optative, imperative, conditional), person agreement with the highest nominative or ergative argument, and the number agreement on T already described. I propose that the syntactic heads bearing these features (e.g., C, T, Asp) fuse morphologically to form a single morphosyntactic node before Vocabulary Insertion. Suffixes like those in (10) compete for insertion into this node. Georgian has an extensive set of screeve suffixes (e.g., see Aronson 1990 for a detailed listing); the following Vocabulary Items are for the aorist tense/aspect:

(10) a. /-es/ $\leftrightarrow$ [Aorist, Group, Class]   3PL.AOR
    b. /-e/ $\leftrightarrow$ [Aorist, Participant]   1/2.AOR
    c. /-a/ $\leftrightarrow$ [Aorist]   3SG.AOR
    d. /-t/ $\leftrightarrow$ [Group] *[To be revised in (13c)]*   PL

Zwicky (1977) argues against the existence of a distinct third-person feature (see also Noyer 1992; McGinnis 2004). In the feature geometry assumed here, third person is represented by the absence of a [Participant] feature. To prevent third-person plural suffixes (like the one in (10a)) from being inserted to indicate agreement with first- or second-person plural arguments, I propose that these suffixes are restricted to third person by the feature [Class]. Since animacy distinctions are reported only in third person, I postulate that the [Class] node is present only for third-person arguments (see Nevins 2002b for a similar analysis of third-person agreement in Arabic).[10] (10a) cannot discharge agreement with a first- or second-person plural argument, since these lack the [Class] feature; (10b) cannot discharge agreement with a third-person argument, since this will lack the [Participant] feature. (10c) ranks below (10a) and (10b) by Pāṇinian disjunctivity. If affixes are inserted from the stem outward (Bobaljik 2000; Embick 2010), we must conclude that (10c) also applies before (10d) (see (5b)). This ordering is predicted if the interpretable features of a node are discharged as soon as possible. Tense/Aspect/Mood features of the TAM node, such as [Aorist], are interpretable, while phi-features such as [Group] are uninterpretable, and thus, by hypothesis, discharged later.

Note that under the proposed analysis, the plural *-t* discharges features of the TAM node. Since this suffix follows the traditional screeve suffixes, rather than being in complementary distribution with them, I postulate that the TAM node is subject to fission. I assume that fission operates as described in Halle 1997a as well as González-Poot and McGinnis 2006: when the insertion of a Vocabulary Item discharges some features of a node, the remaining features fission off to a subsidiary node. Lower-ranked Vocabulary Items then compete for insertion into this subsidiary node. An inserted item discharges features of the subsidiary node, and any features still remaining to be discharged fission off to a new subsidiary node. Evidence for a fission analysis of Georgian TAM suffixes is presented in section 3.3.1. The effect of fission is illustrated in (11). Third-person TAM suffixes in Georgian generally have a special plural form, which discharges a [Group] feature associated with the highest nominative or ergative third-person argument, as in (3a), repeated here as (11a). First- and second-person TAM suffixes lack a number distinction, so [Group], if present, fissions off to be realized as *-t*, as in (11b) (cf. (6b)).

(11) a. g-nax-<u>es</u>
      2.DAT-see-AOR.3PL
      'They saw you (SG/PL).'
   b. v-nax-<u>e</u>-<u>t</u>
      1-see-AOR.PART-PL
      'We saw him/her/it/them.'

In section 3.3.1 I argue that another Vocabulary Item also competes for insertion into the (fissioned) TAM suffix. This is a default number suffix *-s*, illustrated in (12).[11]

(12) g-nax-o-s
     2.DAT-see-OPT-#
     '. . . that he/she see you (SG)'

This *-s* occurs in the Present, Future, Conjunctive, and Optative. I hypothesize that these TAM types share a feature, which restricts the distribution of *-s*. I leave the investigation of this feature for future research; here I simply label it as [F_{TAM}]. The list of TAM Vocabulary Items below, used in the Optative mood, includes the *-s* suffix (13e)[12].

(13) a. /-o/ ↔ [F_{TAM}, Optative]   OPT
     b. /-n/ ↔ [#, Group, Class] in env. [Optative]   3PL.OPT
     c. /-t/ ↔ [#, Group] *[Revised from (10d)]*   PL
     d. /-Ø/ ↔ [#] in env. [F_{TAM}, Participant]   1/2SG.OPT
     e. /-s/ ↔ [#] in env. [F_{TAM}]   3SG/default #

The Vocabulary entry for *-t* in (10d) has been revised in (13c) to include the feature [#], which allows it to block the default number suffix *-s*. Since the third-person plural item in (13b) also blocks *-s*, a [#] feature is included in this item as well. I also postulate a null [Participant] suffix with the feature [#], since *-s* is also blocked when the TAM node agrees in person with a Participant argument. (13a) is correctly predicted to be inserted first if interpretable features are discharged as soon as possible (cf. (10c), which unlike (13a) is blocked by first- or second-person agreement); note that the interpretable features of (13b,d,e) are contextual features only, which are not discharged by inserting these items. The remaining items are inserted after (13a), and are ranked by Pāṇinian disjunctivity. Again, feature discharge by intrinsic features takes the leading role in the ranking of Vocabulary Items, so (13c) ranks above (13d); however, contextual features also play a role, so (13d) ranks above (13e).

Armed with the theoretical tools sketched above, we can now illustrate the operation of fission during Vocabulary Insertion into the TAM node. For example, (14) shows the insertion of Vocabulary Items into the TAM node in (12) *g-nax-o-s*. In (14a), the item /-o/ is inserted. Its F_{TAM} features discharge the corresponding features of the TAM node (indicated as struckthrough), and the remaining features fission onto a subsidiary node, as shown in (14b). (14b) also shows the insertion of /-s/, whose [#] feature discharges the [#] feature of the fissioned TAM node. Presumably, the remaining feature [π] fissions onto an additional subsidiary node; I postulate the insertion of a featureless null default Vocabulary Item into this node, as shown in (14c).

(14) a.   b.   c.

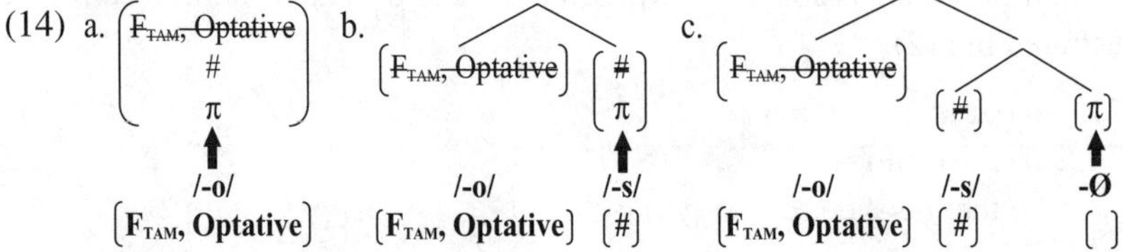

Note that the list of Vocabulary Items in (13) postulates that $[F_{TAM}]$ is merely a contextual feature of *-s*; it is the optative suffix *-o* that actually discharges $[F_{TAM}]$ from the TAM node. This assumption is crucial, since it ensures that plural *-t* ranks above *-s* in the competition for insertion. As discussed in section 3.3.1, this analysis straightforwardly accounts for a heretofore puzzling complementarity between *-t* and *-s*, illustrated in (15). The mechanism by which *-t* blocks *-s* is shown in (16). In (16a), the optative suffix is inserted, as in (14a). In (16), however, the TAM node agrees with a plural argument. The plural suffix *-t* is inserted in (16b), since it discharges more features of the node than *-s* does. The [#] feature is also discharged by *-t*, so *-s* cannot be inserted; only the null default item can be (16c).

(15) g-nax-o-(*-s)-t(*-s)
      2.DAT-see-OPT(*-#)-PL(*-#)
      '. . . that he/she see you (PL).'

(16) a.   b.   c.

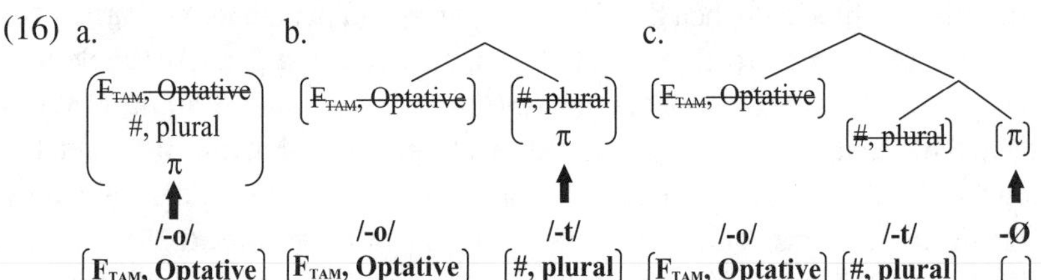

The analysis proposed above has several key components: (i) syntactic competition for agreement with an uninterpretable number feature on T, specified as plural; (ii) clitic movement of first- and second-person pronouns to TP; (iii) encoding of first-person plural in Georgian via a collective first-person feature, [Multispeaker], along with Impoverishment of the plural feature [Group] from dative first-person arguments at the end of the *v*P phase; (iv) morphological fusion of syntactic heads to form the TAM head, which is subject to fission during Vocabulary Insertion; and (v) competition among Vocabulary Items, with items ranked by Pāṇinian disjunctivity applying preferentially to their intrinsic features, along with a requirement to discharge interpretable features as soon as possible. The next section briefly compares this analysis with some important alternatives.

3.3 Evidence for the Proposed Analysis

Although the analysis proposed here is framed within the theory of Distributed Morphology and draws on previous DM analyses of Georgian agreement (Halle and Marantz 1993; Béjar 2003), it also incorporates some insights from Anderson's (1984) analysis, framed within an Extended Word and Paradigm approach, subsequently known as A-Morphous Morphology (Anderson 1992).

In section 3.3.1, I argue for an analysis of Georgian number-marking suffixes that involves Béjar-style syntactic competition, along with morphological competition among Vocabulary Items ranked by Pāṇinian disjunctivity and inserted into positions corresponding to semantically coherent morphosyntactic nodes. This analysis contrasts with Anderson's, which relies on extrinsic ordering of word-formation rules within arbitrary rule blocks. On the other hand, the proposed analysis does adopt Anderson's view that plural *-t* and the number-agreement suffix *-s* occupy the same morphological domain as the TAM (screeve) suffixes, and compete with them for morphological realization. Thus, it differs from the proposals of Nash (1994) and Lomashvili and Harley (2011), who associate the TAM suffixes and *-t* with distinct morphological domains, corresponding to tense and agreement respectively.

In section 3.3.2, I argue that dative 'me' and 'us' in Georgian are distinguished via person, not number, and that Impoverishment of the [Group] feature from a first-person dative argument (associated with *gv-*) prevents it from licensing plural agreement (*-t*). This analysis also contrasts with Anderson's, which seeks to capture the interaction between *gv-* and *-t* by applying the Elsewhere Condition across word-formation rule blocks. However, it adopts Anderson's view that the person prefixes belong to a separate morphological domain from the number suffixes. It thus departs from Halle and Marantz's proposal that the plural feature underlying *-t* fissions off from the person-marking prefix.

3.3.1 Evidence for Specified Agree and Fission in Georgian Number Marking

Examples like those below provide evidence for the analysis presented in section 3.2, that interactions among Georgian number suffixes arise from syntactic competition for a single number-agreement node on T, which fissions during Vocabulary Insertion. (2), repeated here as (17a), shows the unmarked aorist suffix *-a* followed by plural *-t*, while (3b), repeated as (17b), shows the third-person plural aorist suffix *-es*, which blocks *-t*. (18) shows the optative mood, which has an overt TAM suffix *-o* and allows the default number-marker *-s*, as shown in (12), repeated as (18a). However, plural *-t* blocks this *-s*, as shown in (15), repeated as (18b).

(17) a. g-nax-a-t

     2.DAT-see-AOR-PL

     'He/she saw you (PL).'

  b. g-nax-es(*-t)

     2.DAT-see-AOR.3PL(*-PL)

     They saw you (SG/PL).'

(18) a. g-nax-o-s

     2.DAT-see-OPT-#

     '. . . that he/she see you (SG)'

  b. g-nax-o-(*-s)-t(*-s)

     2.DAT-see-OPT(*-#)-PL(*-#)

     '. . . that he/she see you (PL)'

Anderson (1984) accounts for these blocking relations among *-t*, *-s*, and the third-plural TAM suffixes by assigning them to the same word-formation rule (WFR) block. WFRs in the same rule block are in complementary distribution, so only one of these suffixes can appear in any given form. Like Distributed Morphology, Anderson's Extended Word and Paradigm (A-Morphous Morphology) framework assumes that morphology is inserted after the syntactic derivation. The WFR chosen from a given rule block is the highest-ranked one that matches the features provided by the syntax. The ranking is determined in part by the Elsewhere Condition (Pāṇinian disjunctivity), and in part by stipulation.

One difficulty for Anderson's analysis is that the plural *-t* generally follows the TAM suffixes, as illustrated in (17a) and (18b). The only exception is the third-person plural TAM suffixes, which block *-t*, as in (17b). The remaining TAM suffixes agree with third-person singular, first-person, and second-person arguments. These TAM suffixes, which can co-occur with *-t*, would have to occupy a separate WFR block from the third-person plural TAM suffixes. This analysis misses the generalization that the TAM suffixes themselves appear to be syntactically conditioned allomorphs of a single morphosyntactic category;

they covary with tense, aspect, mood, and verb class, and thus are traditionally represented within the same "screeve" or paradigm.

By contrast, the analysis proposed here is in keeping with the view that the domains within which morphological disjunctivity obtains are not arbitrary WFR blocks, but instead are determined by the syntax and by quasi-syntactic operations, such as fusion of functional heads. In other words, since the TAM suffixes are all associated with the same syntactic node, they must compete for insertion within the same morphological domain.

This view has led some scholars (such as Nash 1994; Lomashvili and Harley 2011) to claim that the TAM suffixes occupy a distinct morphological domain from plural -*t*, since they are generally not in complementary distribution with it, as illustrated in (17a) and (18b). However, this approach also faces a difficulty. If the plural -*t* is independent from the TAM suffixes, there is no evident reason for the complementarity between third-person plural TAM suffixes and -*t*. Lomashvili and Harley propose an Impoverishment operation that deletes the plural feature from second-person datives in the context of a third-person plural argument; this prevents -*t* from discharging a plural feature from the second-person argument, ruling out **g-nax-es-t* 'They saw you (PL).' However, the Impoverishment analysis does not rule out multiple exponence of a single plural feature as in **g-nax-es-t* 'They saw you (SG)', with -*es* and -*t* both showing agreement with the third-person subject.[13] Third-person subjects are certainly capable of triggering plural -*t*—where there is no plural form of the TAM suffix to block it, as seen in (19) for third-person dative experiencer subjects.

(19) a. am Kac-s   megobr-eb-i   h-q'av-s
         this man-DAT   friend-PL-NOM   3OBJ-have-#
         'This man has friends.'

    b. am Kac-eb-s   megobar-i   h-q'av-t̲
         this man-PL-DAT friend-NOM   3OBJ-have-PL
         'These men have a friend.'   (Based on Nash 1994, 169)

A fission analysis makes it possible to have it both ways—that is, to keep all TAM suffixes within the same morphological domain, and also to keep -*es* and -*t* within the same morphological domain. In a fission analysis, items competing for insertion are not in complementary distribution. Instead, the highest-ranked compatible item is inserted first, discharging some features of the node, and any remaining features fission off; thus, a lower-ranked item can also be inserted, if it matches the fissioned features. This was illustrated for -*o-t* in (16). This analysis correctly rules out multiple exponence of the same plural feature by both the third-person plural TAM suffix and plural -*t*: once the plural feature is discharged by the TAM suffix it is no longer available to be discharged by plural -*t*. On the other hand, fission alone cannot rule out the combination of two plural suffixes to discharge two separate plural agreement features. Instead, the claim that arguments compete to agree with a single uninterpretable [#] feature on T makes it impossible to obtain two separate plural agreement features on a single T node. The selective nature of syntactic Agree thus rules out *g-nax-es-t 'They saw you (PL).'

The same analysis also accounts for the complementarity of -t and -s, illustrated in (18b). Anderson accounts for this interaction by extrinsically ordering -t above -s. Anderson's analysis correctly predicts the interaction, but only by stipulation; in the analysis proposed here, it arises from two independently justified mechanisms. The first is syntactic competition for agreement. In the proposed analysis, syntactic competition for Agree rules out the possibility of multiple [#] agreement features on T, preventing -t and -s from being inserted to discharge distinct number agreement features of the same TAM node. The second mechanism is Pāṇinian disjunctivity, which ranks -t above -s because the intrinsic features of -s are a proper subset of those of -t (20). Thus, -t blocks -s, not vice versa. In order for Pāṇinian disjunctivity to rank (20a) above (20b), it is necessary to treat the TAM restrictions on -s as contextual, not intrinsic, features. This analysis is justified by the fact that -s can combine with an overt TAM suffix, as in (18a).

(20) a. /-t/ ↔ [#, Group]   pl
     b. /-s/ ↔ [#] in env. [F_{TAM}]   3SG/default #

By contrast, Anderson's analysis does not distinguish contextual from intrinsic features, so Pāṇinian disjunctivity cannot account for the ranking of -t above -s. Indeed, Anderson's (1984, 12) WFR for -s, specified for [−me, −you [. . .]], is notably more complex than the one for -t, which is specified only for [. . .+pl. . .].

3.3.2 Evidence for a Collective Singular Dative First Person in Georgian

As mentioned in section 3.1, there is apparent counterevidence to the syntactic competition analysis—namely, forms like (21), which appear to agree simultaneously with two plural arguments. Indeed, Lomashvili and Harley (2011) explicitly present such forms as evidence against Béjar's (2003) claim that Georgian clauses have only one number-agreement node. Nevertheless, both language-internal and crosslinguistic evidence supports an alternative account of (21) that makes it possible to maintain a syntactic competition analysis of Georgian number agreement.

(21) gv-nax-e-t
     MULTISP.DAT-see-AOR.PART-PL
     'You (PL) saw us.'

The first observation of note is that, while a nominative first-person argument can trigger plural *-t* (22), its dative counterpart is instead associated with a special prefix, *gv-* (23).

(22) a. v-nax-e
      1.NOM-see-AOR.PART
      'I saw him/her/it.'
  b. v-nax-e-t
      1.NOM-see-AOR.PART-PL
      'We saw him/her/it.'

(23) a. m-nax-a
      1.DAT-see-AOR
      'He/she saw me.'
  b. gv-nax-a(*-t)
      MULTISP.DAT-see-AOR (*-PL)
      'He/she saw us.'

Halle and Marantz (1994) propose that the plural *-t* and the person-marking prefix compete for insertion into the same node, bearing the fused phi-features of the subject and object. This node can fission into a prefix and a suffix. On this analysis, the more specific *gv-* wins the competition to discharge the first-person plural feature in (23b), blocking the insertion of *-t*. Lomashvili and Harley (2011) give a similar analysis of this interaction, postulating a morphological template with two agreement positions instead of a fission operation. By contrast, Anderson (1984) treats the prefix and suffix as separate domains (WFR blocks). Nevertheless, he proposes that the Elsewhere Condition applies across WFR blocks. Thus, he proposes that the WFR inserting plural *-t* is blocked by the application of the more specific WFR inserting *gv-*.

Anderson notes that *gv-* only blocks *-t* when there is only a single plural argument. When the dative first-person 'us' co-occurs with a plural second-person argument, both affixes appear (21). Nevertheless, since his theory makes no reference to feature discharge, it provides no account of such forms; it incorrectly predicts that *gv-* will always block the insertion of *-t*. By contrast, on Halle and Marantz's analysis (and Lomashvili and Harley's), disjunctive blocking is tied to feature discharge. In (23b), *gv-* discharges the plural feature of the first-person dative argument, leaving no plural feature for *-t* to discharge. In (21), the second-person argument contributes an additional plural feature, which *-t* can and does discharge.

Previous analyses of the interaction between *gv-* and *-t* assume that both plural arguments trigger number agreement in (21). Such an approach is inconsistent with the analysis proposed here, which crucially maintains that only one argument in the clause triggers number agreement. As it turns out, however, there is independent evidence supporting the proposal made here that dative 'us' in Georgian is represented morphologically as a collective singular first person, rather than as a plural. First of all, the fission analysis of Halle and Marantz (1993) postulates "long-distance" fission, with the two positions of exponence separated by the stem. However, González-Poot and McGinnis (2006) argue that fission must be strictly local. For one thing, local fission yields ordinary binary-branching fissioned nodes, while long-distance fission would create syntactically exceptional ternary-branching structures, with the subsidiary node on the opposite side of the stem, as in (24).[14]

(24)

$$[F_1, F_2, F_3] \quad \text{Stem} \quad \rightarrow \quad [F_1, F_2] \quad \text{Stem} \quad [F_3]$$

Perhaps the most important evidence against a fission analysis is that the core cases of fission split off plural features in second and third persons, but not in first. This pattern is found not only in Georgian dative agreement, but also in Semitic languages (Noyer 1992; Halle 1997a; Tourabi 2002; Nevins 2002a), in Basque (Arregi 2001), and in Kadiwéu (Nevins and Sandalo 2011). In each case, first person distinguishes plurality in the prefix, with no suffix, while second and third person distinguish number in the suffix, with no change to the prefix. Furthermore, ergative agreement in Yucatec Mayan shows a nearly identical distribution to that just described, yet does not readily submit to a fission analysis. In Yucatec, the interaction in question arises between the auxiliary suffix and the verb suffix. The examples in (25) illustrate the observation that the first-person plurality contrast is expressed by the auxiliary suffix (*-in/-k*), which generally indicates person distinctions, while second- and third-person plurality distinctions are expressed instead by a suffix on the verb (*-éeʃ/-oʔobʔ*). These suffixes cannot be sisters of the same stem; they are not even part of the same morphophonological word.[15] In short, Yucatec strongly suggests that long-distance interactions between person and number affixes do not arise from fission.

(25) a. k-in w-áant-ik   'I help him/her.'
      b. k-k   ʔáant-ik   'We help him/her.'
      c. k-a   w-áant-ik   'You (SG) help him/her.'
      d. k-a   w-áant-ik-éeʃ   'You (PL) help him/her.'
      e. k-u   y-áant-ik   'She/he helps him/her.'
      f. k-u   y-áant-ik-oʔobʔ   'They help him/her.'   (Blair and Salas 1995)

Instead, González-Poot and McGinnis propose that such interactions arise from the featural representation of 'us', following Ritter 1997, Harley and

Ritter 2002, and Nevins 2002b. Noting that some pronoun systems distinguish plurality only in first person, Harley and Ritter propose that person, rather than number, specifications can distinguish first-person singular and plural; in such systems, 'us' is effectively a collective singular category, not plural. Ritter (1997) and Nevins (2002b) also postulate a person distinction between first-person singular and plural in Semitic languages, where number is expressed in second and third persons.

This type of analysis makes it possible to account for the interaction between *gv-* and *-t* in (23b) without recourse to fission or a templatic analysis. If 'us' is a collective singular form in the dative, rather than plural, then both the special first-person prefix and the absence of the plural suffix are expected. The same analysis can be given for the apparent double-plural case in (21). If the first-person dative argument lacks a plural feature, it will not trigger plural agreement on T. Another argument can do so instead, as in (21), where a second-person plural argument triggers agreement, or in (26), where a third-person argument does.

(26) gv-nax-es

    MULTISP.DAT-see-AOR.3PL

    'They saw us.'

One question that Georgian raises is how to restrict the person-based plurality distinction to the dative case, since nominative first-person arguments do trigger plural agreement, as shown in (22). On the analysis I have proposed, both dative and nominative first-person pronouns have the same syntactic representations, including both the person feature [Multispeaker] and the number feature [Group]. The [Group] feature is deleted from dative pronouns by postsyntactic morphological Impoverishment. Impoverishment must apply after the dative case is valued, but before the number-agreement feature on T has access to the [Group] feature of the dative pronoun; thus I propose that it applies at the spell-out of the *v*P phase. Nominative first-person pronouns can bear the [Multispeaker] feature, but there happens to be no nominative agreement item that specifies this feature, so the same prefix is used for singular and plural. However, the plural suffix does distinguish between singular and plural nominative first-person agreement.

This analysis predicts that there are languages with both [Multispeaker] and [Group] in which Impoverishment does not apply, and in which the Vocabulary Items express the range of possible person and number distinctions. In such a case, first-person plural would be marked with both a special first-person item (specified for [Multispeaker]), and a plural suffix (specified for [Group]). This prediction may be correct, as illustrated in (27), which shows pronouns in

Nama (Hagman 1977, 44, cited in Harley and Ritter 2002). In this language, first-person dual and plural pronouns have a special person-marking stem *sií*, which contrasts with first-person singular *tií(ta)* as well as with second-person/ inclusive *saá* and third-person *//iĩ*.[16] They also bear [Group]-marking suffixes, including *-m̀/-m* in the dual (which includes both [Minimal] and [Group] specifications), and *-e* in the plural.

(27) *Nama pronouns*

	SINGULAR		DUAL		PLURAL	
	FEM	MASC	FEM	MASC	FEM	MASC
1st	tiíta	tiíta	siím̀	siíkxm̀	siíse	siíke
incl	—	—	saám	saákxm̀	saáse	saáke
2nd	saás	saáts	saárò	saákxò	saásò	saákò
3rd	//'iĩs	//'iĩp	//'iĩrà	//'iĩkxà	//'iĩti	//'iĩku

I conclude that the proposed analysis is viable. Both [Multispeaker] and [Group] are present on first-person plural representations in Georgian, but [Group] is deleted from these representations in the dative, so only [Multispeaker] remains to make the singular/plural distinction. In short, there is no double plural marking in Georgian; only one argument agrees with the [Group] feature on T. Thus, there are no counterexamples to the generalization that only one argument per clause can trigger number agreement in this language.

3.4   Conclusions

This chapter argues that the facts of Georgian number agreement are insightfully captured by an analysis involving two main components: syntactic competition for agreement with a T node specified for uninterpretable plural number features; and Vocabulary-driven fission of a node expressing Tense/ Aspect/Mood (TAM) and phi-agreement. Competition for number agreement ensures that only one (preferably, plural) subject or clitic object triggers number agreement, while fission ensures that the default plural suffix *-t* expresses plural agreement if and only if the TAM (screeve) suffix has no special plural form. A third, and equally essential, component of the analysis is the claim that Georgian first-person plural arguments bear a special person feature, [Multispeaker], which co-occurs with the plural feature [Group]; and that the [Group] feature is deleted from dative first-person arguments, so that these fail to trigger plural number agreement, and instead are distinguished from singulars purely by the [Multispeaker] feature. This component of the analysis accounts for the fact that Georgian dative first-person pronouns trigger a special person-marking prefix (*gv-*) and no plural suffix (*-t*); it also accounts for the apparent double plural marking in forms with *gv-* and a second- or third-person plural nominative argument. According to the syntactic competition analysis, such double number marking should be impossible, and, indeed, if *gv-* marks [Multispeaker] rather than plural number, these forms actually do not involve double plural marking. The proposed analysis also distinguishes between intrinsic features of a Vocabulary Item—those discharged by the item—and contextual features, which are not discharged. This distinction makes it possible to account for the capacity of plural *-t* to block a default number suffix *-s*. It is demonstrated that *-s* does not discharge the TAM features conditioning its insertion, so these are contextual features, and its intrinsic features rank below those of *-t* by Pāṇinian disjunctivity. I have argued that the independently motivated proposals made here successfully address the empirical and theoretical challenges faced by previous analyses. Thus, this chapter provides the most comprehensive analysis of Georgian number agreement to date. Moreover, it does so by maintaining Halle and Marantz's foundational insight that morphological disjunctivity obtains within coherent domains, corresponding to individual or fused syntactic nodes.

Notes

Thanks to Alec Marantz, Léa Nash, Heidi Harley, and Ora Matushansky for valuable discussion of Georgian inflectional marking, and to two anonymous reviewers of this chapter for their tremendously helpful feedback. Thanks also to Léa Nash for her Georgian judgments.

1. It may be that this second probe operation is possible because an unmatched specified feature deletes after probing (Béjar 2003), effectively creating a new probe from the remaining (general) feature.

2. As a reviewer notes, it is unclear why unmatched specified features should be able to delete, if unmatched general features instead cause a derivation to crash. I leave this matter for further investigation.

3. Béjar and Rezac (2003a) propose that the directionality of Agree follows from locality, and from the assumption that the probe is the root of the tree, not the head of the root projection. An agreeing root probes downward as soon as its head is merged, since that is its local domain. As the tree extends upward, any newly merged argument is closer to the (new) probe than any lower argument.

4. Bruening (2001) provides binding evidence for a similar phenomenon in Passamaquoddy, which also separates proximate from obviative third-person arguments.

5. Another possibility, suggested by a reviewer, is that nonclitic objects are separated from T by a phase boundary, and are inaccessible for this reason. Both accounts are also consistent with the fact that [Participant] nominative objects trigger person agreement on the TAM suffix, since these objects are clitics.

6. One puzzle for the proposed analysis is that a third-person object in Georgian can A-scramble over the subject (McGinnis 1998, 2004)—yet, unlike clitic movement, this movement does not feed number agreement.

7. I assume that, in languages with active [Group] and [Multispeaker] features, all [Speaker] representations bearing the feature [Multispeaker] also bear the feature [Group], and vice versa. This constraint, plausibly motivated by semantic considerations, rules out improbable systems with a four-way number distinction in first person.

8. See Harley and Tubino Blanco, chapter 7, this volume, for additional evidence that the vP phase plays a role in the morphology.

9. There is evidence that the π-agreement node is an independent site for Vocabulary Insertion, distinct from the v node itself, whose morphological realization is discussed by Nash (1995). I leave this issue aside here.

10. Evidence for this claim (and moreover that the [Class] feature of (10a) is further specified as [Animate]—or rather [Sentient], following Hanson 2003 and Bliss 2005) is that only sentient subjects actually trigger plural TAM agreement in Georgian (Harris 1981, 149; Sedighi 2005).

11. This -*s* is not a third-person singular suffix, since in clauses with a dative subject it can occur with either a singular or plural nominative third-person object—or with no object (Marantz 1989).

12. To be more precise, -*o-n*/-*o* are the optative TAM suffixes for regular first-conjugation verbs and regular second-conjugation verbs in *i*- (Aronson 1990, 142). Regular second-conjugation verbs in -*d*- take -*a-n*/-*a* instead.

13. McGinnis (1996) proposes a morphophonological rule that deletes plural -*t* after a plural suffix. This analysis does correctly rule out multiple exponence in **g-nax-es-t*; however, it does not predict the complementarity between -*t* and -*s*, shown in (18b) and discussed below.

14. Lomashvili and Harley's (2011) templatic analysis of Georgian agreement avoids postulating ternary-branching nodes, though at the cost of not clarifying exactly how syntactic agreement maps onto morphosyntactic template slots.

15. A general rule in Yucatec inserts a glottal stop to repair an onsetless syllable, as in *ʔáant-ik* (25b). It is evident that the verb prefix also reflects person distinctions and the first-person number distinction. A fission analysis could account for the interaction between verb prefixes and suffixes, but would leave the auxiliary suffixes unexplained; even concord between the auxiliary suffix and a fissioned verb prefix would not automatically predict that both affixes would happen to have a special first-person plural form. On the other hand, if first-person plural is a separate person category, the alternation is expected in both person-marking positions.

16. // is a lateral click.

4 More or Better: On the Derivation of Synthetic Comparatives and Superlatives in English

Ora Matushansky

4.1 Introduction

As illustrated in (1), English comparatives and superlatives can be synthetic, derived with the suffixes *-er* and *-st*, respectively, or analytic, requiring the freestanding morphemes *more* and *most*. While in some syntactic environments, such as metalinguistic comparison (see Bresnan 1973 and Kennedy 1999, among others), only analytic forms are possible, generally only "short" adjectives allow synthetic forms:

(1) a. smarter, tallest, simplest, shallower . . .

    b. most intelligent, more prudent, most splendid, more beautiful . . .

It is a standard assumption (see, e.g., Emonds 1976), which I also adopt here, that there is no interpretational difference between the bound morphemes *-er* and *-st* on the one hand, and the free morphemes *more* and *most* on the other. Traditionally (Corver 1997b), synthetic forms have been derived by the movement of A° to Deg°, with analytic forms arising from the insertion of the support morpheme *much* when head movement fails (*much*-support). Recently, however, an alternative proposal has relegated the derivation of synthetic comparatives and superlatives to a postsyntactic lowering operation: either Local Dislocation (Embick and Noyer 1999, 2001; Embick 2007a) or Morphological Merger (Bobaljik 2012). The derivation of synthetic forms by Affix Hopping has not been proposed.[1]

In this chapter I argue against lowering/postsyntactic approaches to the derivation of synthetic comparatives and superlatives by demonstrating that the evidence against the head-movement analysis adduced by Embick and Noyer is nondecisive and that a postsyntactic approach cannot account for the finer details of the distribution of synthetic and analytic forms.

4.2 Against Local Dislocation

As is well known, the formation of English synthetic comparatives and superlatives is subject to a prosodic constraint (Marantz 1988; Pesetsky 1979, 1985; Quirk, Greenbaum, Leech, and Svartvik 1985; Sproat 1985): the *-er/-est* suffixes can only attach to one-foot stems (McCarthy and Prince 1993).[2] In other words, only monosyllabic adjectives and disyllabic adjectives with a light second syllable (e.g., *silly–sillier*, *yellow–yellower*) can give rise to synthetic forms:

(2) a. smarter, #more smart; brightest, #most bright
    b. *beautifuller, ✓ more beautiful; *intelligentest, ✓ most intelligent

Embick and Noyer 1999, 2001 argue that deriving synthetic forms by head movement is incompatible with the "Late Insertion" hypothesis, according to which lexical roots are not present in syntax, but are inserted after the spell-out (Marantz 1994): since head movement occurs before Vocabulary Insertion, no effect from the choice of the lexical root is expected. Embick 2007a argues that the problem also extends to "early insertion" frameworks: syntax should not be sensitive to phonological properties of particular lexical items. Conversely, Local Dislocation, a postsyntactic operation applying to linearized structures, can easily be made sensitive to the phonological properties of the adjectival stem:[3]

(3) Local Dislocation rule for comparatives and superlatives
    (Embick 2007a, 25)[4]
    Deg[CMPR,SUP]⌢[. . .X. . .]$_a$ → [. . .X. . .]$_a$⊕Deg[CMPR,SUP]
    (in English: where the phonological form of [. . .X. . .]$_a$ meets the
    relevant prosodic condition)

The core property distinguishing Local Dislocation from both head movement and Affix Hopping is that the former occurs at or after Vocabulary Insertion. As a result, only Local Dislocation can be sensitive to the prosodic structure of individual lexical items. However, as argued by Bobaljik 2012, the problem is that Local Dislocation cannot deal with suppletion: cross-linguistically, synthetic comparatives and superlatives of adjectives such as *good*, *bad*, *little* and *many/much* are often suppletive, and English is obviously no exception:

(4) a. good → better, best
    b. bad → worse, worst
    c. little → less, least
    d. many/much → more, most

Since the Local Dislocation rule in (3) contains a reference to the phonological form of the adjective in question, the adjectival stem must be spelled out before combining with the comparative/superlative suffix, which incorrectly predicts that stem suppletion, as in (4a), should be impossible. To avoid this outcome, it could be suggested that Vocabulary Insertion into the complex head [a v]$_a$ is conditioned by the presence of a comparative/superlative morpheme in the same maximal projection. The empirical problem with such a solution is obvious when we realize that the interaction between the choice of the analytic or the synthetic form and the availability of suppletion should give rise to four options, of which is missing: precisely the one that is enabled if Vocabulary Insertion can be conditioned from outside the target head:

(5) a.   intelligent → more/most intelligent
    b.   cute → cuter/cutest
    c.   *wuggal → more/most galliwug
    d.   good → better, best

Since the pattern in (5c) is crosslinguistically not attested, Bobaljik 2012 argues that the derivation of synthetic forms must precede Vocabulary Insertion and is therefore achieved by Morphological Merger or head movement. To constrain this process to only apply to certain roots, Bobaljik proposes that it is triggered by the diacritic feature [+m] on the root node, where only roots marked [+m] can be inserted; similarly, Graziano-King 1999 suggests that the selection of the synthetic form is listed in the lexicon. To explain McCarthy and Prince's prosodic generalization, Bobaljik 2012 hypothesizes that the assignment of the diacritic generalizes over statistical regularities in the input of the language learner.

Setting aside the stipulative nature of this proposal, excluding the phonological form of the adjective from the derivation of synthetic forms seems to be incorrect. As demonstrated in Mondorf 2009, 24–30, the final consonant cluster of the stem affects preferences in cases of apparent free variation. Thus, considerations of euphony explain why adjectives ending in *-st* are highly unlikely to form synthetic superlatives (cf. Jespersen 1956), while adjectives ending in *-er* or *-re* show a strong tendency for analytic comparatives (cf. also Plag 1998). The diacritic-feature approach does not lead us to expect such preferences where the synthetic form is grammatical, and the differing behavior of comparatives and superlatives in function of the final cluster can only be modeled by postulating two diacritic features instead of one.

Bobaljik 2012 provides yet another empirical generalization that can be used to argue against postsyntactic approaches to the formation of synthetic comparatives and superlatives: crosslinguistically, if the comparative of an adjective is suppletive, the corresponding change-of-state verb is also suppletive. If this correlation is due to the fact that deadjectival change-of-state verbs are derived from the comparative rather than the positive form of an adjective, then comparatives must be derived in narrow syntax, in order for deadjectival verbs to be able to undergo such syntactic processes as head movement (albeit not in English) or further derivation, including transitivization.

Countering Embick's objections to the derivation of synthetic forms in narrow syntax, I will now show that head movement can in principle be made sensitive to the choice of a specific lexical item. Following Corver 1997b, let us assume that the comparative/superlative Deg° bears the uninterpretable feature [degree]. Assuming, by an analogy with √-to-v°, that √-to-a° movement is obligatory[5] and the affixal status of the comparative/superlative Deg° triggers overt A°-to-Deg° movement, as in (6a), (6b) results:[6]

(6) a.   b.

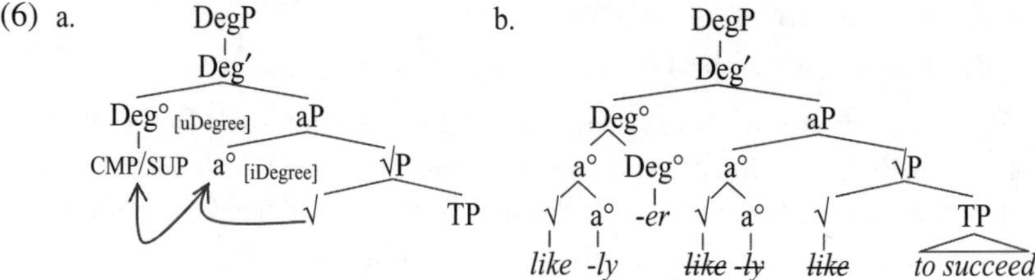

At spell-out the structure in (6b) (Abney 1987; Bowers 1987; Corver 1990, 1991, 1997a, 1997b) is evaluated and the lexical properties of various morphemes come into play. Assuming that *-er/-est* can only attach to "short" stems, a "long" adjectival stem will fail to be merged in this position, which leads to the Last Resort operation of *much*-support. Since the root node and the affixal a° still need to be spelled out, the adjectival stem surfaces in the lower position (a°), yielding the analytic form.[7]

A potential objection comes from the fact that the syntactic structure in (6) is not the only one hypothesized for comparatives and superlatives. Indeed, Bowers 1975, Jackendoff 1977, and more recently, Heim 2000, 2006, and Schwarzschild and Wilkinson 2002, among many others, presuppose that the comparative DegP is merged as [Spec, aP]—a structure that is not compatible with either head movement or Affix Hopping:[8]

(7)

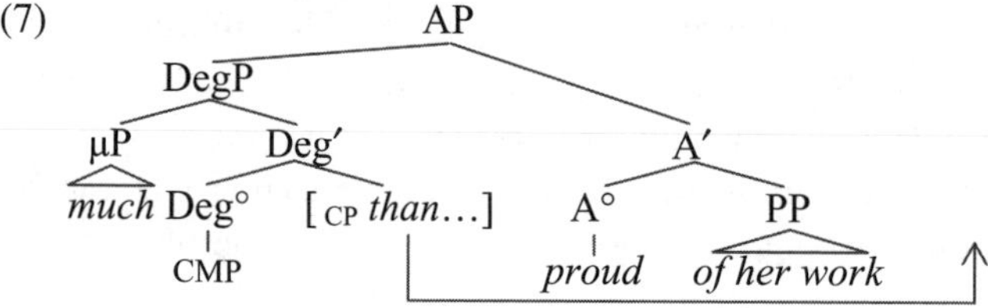

To rule out the structure in (7) it must be demonstrated that synthetic comparatives cannot be derived by either Local Dislocation or Morphological Merger—the two operations that can combine the relevant terminals in this configuration. As discussed above, Local Dislocation is excluded by its inability to account for suppletion (Bobaljik 2012), while Morphological Merger cannot straightforwardly account for the prosodic constraints on the availability of the synthetic forms. In addition, evidence against the derivation of synthetic comparatives and superlatives by any form of lowering, including Affix Hopping, comes from coordinated comparatives like (8), which Jackendoff 2000 argues to be incompatible with the usual assumptions about the structure of comparatives:

(8) a. more and more beautiful
    b. prettier and prettier

Indeed, the coordinated comparative morphemes in (8a) can be projected in the structure in (7)—that is, as a conjunction of DegPs in [Spec, AP] in (9)—or in the structure in (6), as the coordination of Deg° heads in (10):

(9)

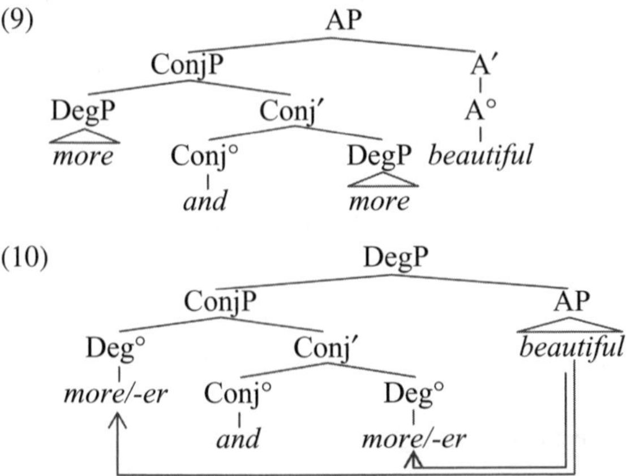

(10)

The question arises how to derive (8b) from these underlying structures. As Jackendoff notes, no lowering operation is able to derive the coordination of synthetic comparatives (8b) in the structures in (9) or (10): an additional copy of the adjective is required for both affixes to combine with a phonological host. Crucially, since head movement and Affix Hopping are inapplicable in the configuration in (9), there is no way to derive examples like (8) and we must conclude that the configuration in (7) is not available, contra Heim 2000, 2006.[9] Noting that head movement in the configuration in (6) would represent an otherwise unattested case of across-the-board insertion (rather than the usual across-the-board extraction), Jackendoff concludes that (8) cannot be derived from the structure in (10) either.

The alternative that Jackendoff does not consider is syntactic reduplication:[10]

(11)

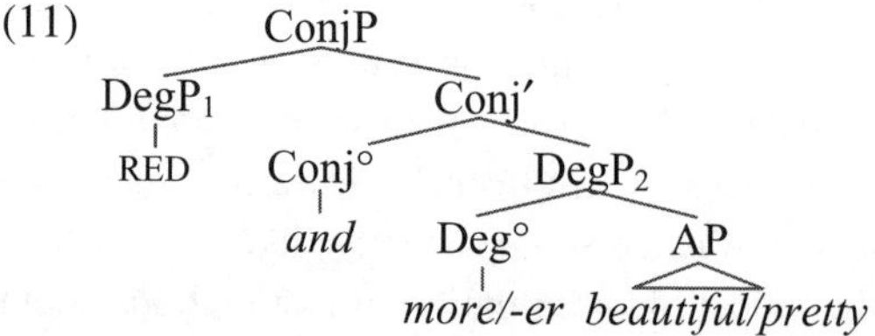

Assuming that RED uniformly copies the contents of Deg°, the structure in (11) derives both (8a), where reduplication results in the repetition of the degree morpheme, and (8b), where reduplication is preceded by head movement of A° into Deg°, yielding a complex head as the source for reduplication. Lowering analyses, on the other hand, cannot take the same syntactic head as the source for the reduplicative morpheme. Since in synthetic forms Deg° is adjoined to A° rather than the other way around, RED has to copy Deg° in analytic forms and A° in synthetic forms. Likewise, allowing RED to copy the first phonological word following *and* seems unlikely for syntactic reduplication.

The final issue to be resolved (noted as a challenge for Local Dislocation in Matushansky 2001 and discussed by Embick 2007a in a footnote)[11] is the comparative bracketing paradox (Pesetsky 1979, 1985; Sproat 1985). As is well known, the addition of the negative prefix *un-* does not affect the ability of an adjectival stem to form a synthetic comparative or superlative:

(12) a. [[un + happy, likely, lucky] + Deg°] → unhappier, unhappiest, unluckiest . . .

    b. [[un + interesting, fortunate] + Deg°] → most uninteresting, more unfortunate . . .

As Pesetsky 1979 shows, for the purposes of synthetic comparative/superlative formation *un*-prefixed adjectives behave as if the resulting structure is (13a) rather than the semantically transparent (13b). To solve this problem Pesetsky 1985 derives the structure in (13b) from that in (13a) by LF movement of the suffix, as in (13c)—a proposal that cannot be adopted if synthetic forms are derived in syntax, as in (14) (see Sproat 1985 and Marantz 1988 for other problems with Pesetsky's analysis):

(13) a. [*un*-[A-*er/st*]]

    b. [[*un*-A]-*er/st*]

    c. [[*un*-[A-~~er/st~~]]-*er/st*]

(14)

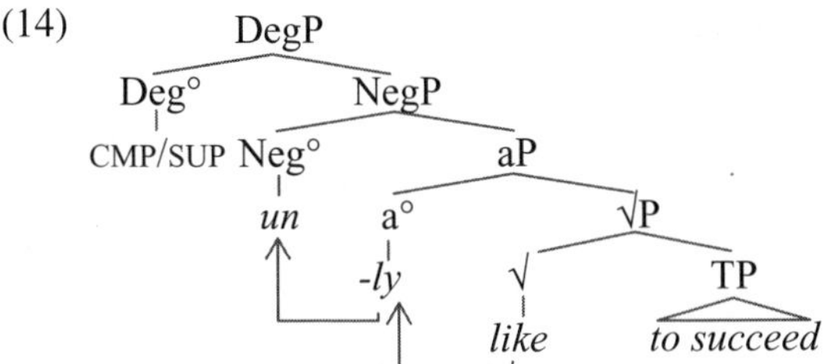

The fact that the negative prefix *un-* does not affect synthetic comparative and superlative formation in the structure (14) can be accounted for under the hypothesis (Siegel 1974; cf. also Pesetsky 1979 on the postcyclic status of Russian prefixes) that the prefix *un-* is postcyclic. If only one-foot adjectives can give rise to synthetic forms (McCarthy and Prince 1993) and a postcyclic prefix is extrametrical, *un-* is correctly predicted to have no effect on the derivation of synthetic forms. The extrametrical status of at least some prefixes[12] is also supported by the exceptions to McCarthy and Prince's generalization— that is, those English disyllabic adjectives with final stress that nonetheless allow synthetic forms. Only for two of these, namely, *diffuse* and *remote*, does the frequency of synthetic comparatives approach the frequency of analytic comparatives (Hilpert 2008): 14 occurrences of *diffuser* to 34 occurrences of *more diffuse* and 87 occurrences of *remoter* to 179 occurrences of *more remote*. Both these adjectives could have been reanalyzed by native speakers as containing a prefix, which would explain why no other stress-final disyllabic adjective approaches these ratios or such frequencies.[13] Whether the postcyclic status of a morpheme is realized as a diacritic or in some other way is orthogonal to my purposes here.[14]

To summarize, comparative and superlative suppletion provides evidence against deriving synthetic forms by Local Dislocation (Bobaljik 2012). Paradoxically, the prosodic constraint on synthetic forms is shown to support the head-movement analysis and to be incompatible with Morphological Merger. All lowering operations, as well as Jackendoff's (1977) structure (7) in general, are excluded by Jackendoff's (2000) conjoined comparatives, which also argue for a head-movement analysis in conjunction with reduplication. Finally, the cyclic approach to the comparative bracketing paradox is equally compatible with the derivation of synthetic forms by head movement or by postsyntactic lowering operations.

In the next section I simultaneously address the blocking effect of modifying adverbs and provide further evidence against a postsyntactic analysis by showing that the derivation of synthetic forms is also conditioned by the semantics of the adjectival stem.

4.3 Scalarity and Norm-Relatedness

As I observed in Matushansky 2001 (the generalization is implicit in the work of Clarke 2001 as well; see also Jovanović 2009), nonscalar adjectives like *French*, *right*, or *male* do not form synthetic comparatives or superlatives in English, irrespective of their phonology:

(15) This is a $^{??}$realer/*goldener/$^{??}$faker/*Frencher sword.

Martin Hilpert's research provides support for this generalization. A measure of inherent scalarity is the ratio of comparatives (or other contexts engaging the degree argument, such as superlatives, combination with *such*, etc.) to positive forms. While Hilpert 2008 observes that higher scalarity leads to a higher percentage of morphological comparative formation, Martin Hilpert (personal communication) notes that the effect is the strongest for short adjectives: analytic comparatives and superlatives of monosyllabic adjectives typically have low scalarity.

Embick 2007a suggests that the unavailability of synthetic forms for nonscalar adjectives is due to their semantic incompatibility with comparison. While this hypothesis explains why nonintersective monosyllabic adjectives (e.g., *main*, *past*, *real*) permit neither synthetic nor analytic forms, as illustrated in (16), it cannot be all of the answer. Indeed, intersective nonscalar adjectives can form analytic comparatives/superlatives with a coerced interpretation (= having more/the most properties associated with being French, right, or male):

(16) a. This is the *mainer/$^{*/??}$more main reason.
     b. He is a *paster/$^{*/??}$more past king.

(17) a. *Becky's aunt is Frencher/deader/wronger than Napoléon.
     b. Becky's aunt is more French/more dead/more wrong than Napoléon.

Two ways of deriving the impossibility of synthetic forms with nonscalar adjectives can be envisaged. Under one view, a nonscalar adjective does not bear the [degree] feature and therefore cannot be attracted to Deg°, as in (18a). Under this view, coercion is not reflected in syntax. Under the other view, coercion is effected by a separate head, as in (18b). To block the derivation of the synthetic form this head must be stipulated to not attract the adjective. A further problem with this latter hypothesis is that additional stipulations must be made to account for the fact that the coerced meaning of nonscalar adjectives is not available in the absence of a degree head.

(18) a.

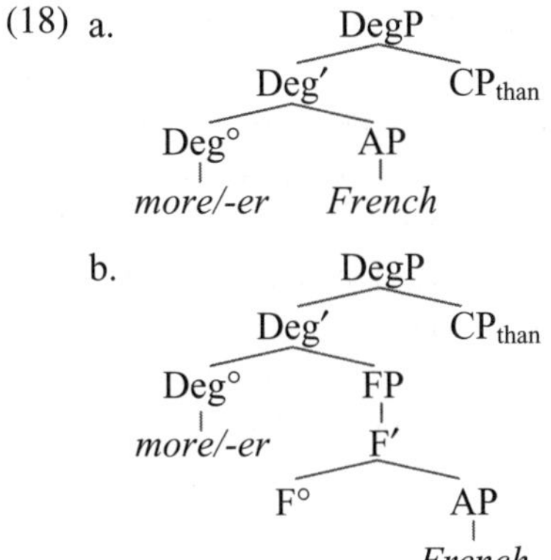

Crucially, such purely semantic features as [degree] are not expected to be available after the spell-out on the PF branch of the computation. The fact that synthetic comparatives and superlatives must be scalar therefore argues against deriving them postsyntactically.

I will now argue that the scalarity constraint on the derivation of synthetic forms correctly predicts that comparatives and superlatives that are formed from APs modified by adverbs (Embick and Noyer 1999, 2001; Embick 2007a) and PPs must be analytic.

4.3.1   The Semantics of Synthetic Forms

In considering the interaction of synthetic comparative/superlative formation with adverbial modification, Embick 2007a discusses three semantic classes of adverbs:

(19) a. Degree

   amazingly smart, incredibly tough, unbelievably short

   b. Manner

   ploddingly slow, happily drunk, clearly glib, flatly honest, rudely late

   c. Aspect

   physically strong, technically proficient, structurally weak

Starting with degree adverbials (Embick's (2007a) "roughly evaluative adverbs"), Embick and Noyer 1999, 2001, and Embick 2007a take the contrast below as an argument against the head-movement derivation of synthetic forms: since verb movement (a paradigmatic example of head movement) is not blocked by structurally intervening adverbs, the question arises why such adverbs should block synthetic comparative/superlative formation:

(20) a.   Mary is the most amazingly smart person.

   b. *Mary is the amazingly smartest person.

Challenging the relevance of (20b), Williams 2006 suggests that the ungrammaticality of such examples is due to the fact that they can only be interpreted as metalinguistic (Bresnan 1973), while Kiparsky 2005 proposes that the freestanding superlative morpheme in (20a) can only be interpreted as forming a constituent with the adverb *amazingly*. While accepting Embick's arguments against both claims, I emphasize that comparatives and superlatives of modified APs necessitate a prior context introducing the relevant concept into the discourse. To exclude Kiparsky's bracketing, consider superlative AP predicates, where the bracketing in (21a) is excluded and (21b) remains the only structure available:[15]

(21) Jessamine was the most amazingly drunk.
    a. *[the most amazingly] drunk
    b.   the most [amazingly drunk]

Crucially, native speakers accept (21) only if the prior context contains some discussion of amazingly drunk individuals, out of which the most [amazingly drunk] one is discussed:

(22) a. We were all amazingly drunk then, but Jessamine was the MOST amazingly drunk.
     b. *Guess what—Ron has become the most amazingly drunk in the bar.

Jonathan Bobaljik (personal communication) suggests that *amazingly drunk* in this context is a complex lexical entry, akin to the syntactically complex structures that can appear as the left member of a compound (e.g., *an "off the beaten track" place, a "holier than thou" attitude*). If correct, this hypothesis would exclude synthetic comparative formation by suggesting that an affix cannot attach to such a structure. The main problem with this proposal is that it is not independently motivated. For instance, concept formation is also operative in DP-internal AP modification (Bouchard 2002, 2005) and is fully compatible with such inflectional morphology as gender, number, or case, as illustrated by Russian:

(23) a. o   belyx   medved'ax
      about white-PL.LOC   bear-PL.LOC
      about polar bears
    b. ot   ussurijskoj   tigr-ic-y
      from   Siberian-FEM.GEN   tiger-FEM-FEM.GEN
      from a Siberian tigress

It has also been claimed that modified NPs are compatible with derivational morphology, leading to such bracketing paradoxes as *the nuclear scientist, a derivational morphologist*, or indeed (23b) (cf. Pesetsky 1979), while Hoeksema 2012 notes that synthetic forms of elative compound adjectives, such as *dirt poor*, are marginally possible in Dutch. The impossibility of synthetic comparatives and superlatives for degree-modified APs cannot therefore be due to the incompatibility of concept formation with affixation, though it might interact with head movement.

Having rejected the head-movement account, Embick and Noyer hypothesize that (20b) cannot be derived because the adverb linearly intervenes between the comparative affix and the adjective, blocking Local Dislocation. However, the intervention effect is not linear. For a degree modifier PP to be interpreted in the scope of the comparative (with the concomitant concept formation, as in (20b)), the analytic form, as in (24), must be used; the synthetic forms in (25) can only mean that the degree to which Jude is smarter than Joe is amazing:

(24) Jude is more smart to an amazing degree than Joe.

(25) a. Jude is smarter to an amazing degree than Joe.
      b. Jude is smarter than Joe to an amazing degree.

To explain these facts I propose that the unavailability of synthetic forms with modified APs is due the fact that, being norm-related, they are not scalar.[16] The link between norm-relatedness and nonscalarity is supported by the fact that analytic comparatives of one-foot scalar adjectives are typically norm-related (Rett 2008):[17]

(26) a. The most clear/clearest evidence comes from the third trial.
      b. The ending is even more subtle/subtler.

Under its most natural interpretation, *the most clear evidence* in (26a) is *clear*, while the truth of (26b) entails that the ending is subtle. Synthetic comparatives and superlatives in the same environment do not have this effect: *the clearest evidence* could still be murky, and an ending that is *subtler* can still be pretty blunt.

Further evidence linking analytic forms to norm-relatedness comes from the attenuative suffix *-ish*. Though adjectives derived with *-ish* are vague (*a tallish mountain* is clearly not of the same height as *a tallish girl*), they do not form synthetic comparatives and superlatives. While it could be suggested that their prosody (they are minimally disyllabic) is to blame, the lexical semantics of *-ish* is a more likely culprit. As argued by Kagan and Alexeyenko 2011, *-ish* asserts that a property holds of an individual to a degree slightly exceeding the standard of comparison (i.e., *-ish* is clearly norm-related).

On the syntactic side, two ways of encoding norm-relatedness have been proposed. Under one view, that of Bierwisch 1989 and Krasikova 2009, 2010, the basic meaning of a scalar adjective is vague and norm-related, while its scalar interpretation is derived. In this approach a norm-related adjective is not marked [degree] and therefore cannot be attracted to Deg°, as in (18a). On the other hand, Rett 2008 adopts the opposite perspective, where the norm-related interpretation results from the introduction of the EVAL operator (intended also as a replacement of POS; see Kennedy 1999). If EVAL is a head that does not attract A°, as in (18b), the derivation of an analytic form is impossible. As a result, under both views norm-related adjectives should pattern with non-scalar adjectives in not giving rise to a synthetic comparative or superlative.

What remains now is to demonstrate that adverbial modifiers yield nonscalar APs. The use of expressive adverbs, for instance, entails the applicability of the positive form:

(27) a. *We are none of us amazingly drunk, but Peter is the most amazingly drunk.

    b. *Neither of them is amazingly drunk, but Peter is more amazingly drunk than Sue.

Both examples (27) give rise to a contradiction on the assumption that a degree-modified AP should entail the positive form. Degree modification in the second conjunct of (27a) must entail that Peter is amazingly drunk, which is denied by the first conjunct; the same is true for (27b), and a contradiction results.[18] The question therefore arises whether the derivation of synthetic forms is blocked by the syntax of norm-relatedness.

I will now demonstrate how the scalarity constraint on synthetic comparative/superlative formation can also account for the blocking effect of non-degree adverbial modifiers.

4.3.2 Manner Adverbials

As noted by Embick 2007a, adverbs specifying the manner in which the property denoted by the adjective holds also block synthetic comparative/superlative formation:

(28) a. Robert is more unobtrusively smart/*unobtrusively smarter than Jessamine.

    b. Lois was more earnestly dull /*earnestly duller than David.

Although many manner adverbs, such as *ploddingly* in *ploddingly slow*, do not have natural PP paraphrases, for less idiomatic modifiers it can be demonstrated that PP adverbials also block synthetic comparative formation: the adverbial PP can be interpreted in the scope of the comparative morpheme in (29a), but not in (29b). More precisely, only the former is compatible with the prior context where the relevant concepts (*smart in an unobtrusive way, dull in an earnest way*) are introduced:[19]

(29) a. Jackie is more smart in an unobtrusive way/dull in an earnest way than Rose.
      b. #Jackie is smarter in an unobtrusive way/duller in an earnest way than Rose.

I contend that this pattern is expected once we take into account the semantic contribution of the adverbial. Thus the fact that manner adverbials can combine with nonscalar adjectives or even with verbs shows that they are not lexically specified for a scalar AP, unlike degree adverbials (examples from COCA (Davies 2008–)):

(30) a. It is an amusing alternative that all but the [most **ploddingly**] **literal-minded** would find unobjectionable for a summer evening's entertainment.
      b. Amidst the six giant turbines, huge orange cranes **ploddingly move back and forth** overhead.

Abstracting away, for the sake of simplicity, from all intensional arguments of the verb would give manner adverbials the semantic type $\langle\langle e, t\rangle, \langle e, t\rangle\rangle$. In order for them to combine with a scalar AP, the AP would have to be interpreted as vague/norm-related (type $\langle e, t\rangle$), as discussed above. As a result, we correctly predict that comparatives of manner-modified APs are norm-related. Assuming that norm-related APs have the syntax of a positive form entails that they can only be analytic:

(31) a. Jackie is more unobtrusively smart than Rose.
          ⇒ Jackie and Rose are both unobtrusively smart.
      b. Lee is more quietly brilliant than Charles.
          ⇒ Lee and Charles are both quietly brilliant.

To summarize, the fact that both manner adverbs and PPs denoting manner disallow the formation of synthetic comparatives and superlatives removes a potential argument in favor of deriving synthetic forms by Local Dislocation. The fact that manner adverbials force the AP they combine with to be interpreted as nonscalar allows us to derive the blocking effect of manner adverbials from the scalarity constraint on the availability of synthetic forms.

4.3.3 Aspect Adverbials

As observed by Embick 2007a, n32, aspect adverbials (cf. Bartsch 1987; Kennedy 1997; Bierwisch 1989) are compatible with synthetic comparatives or superlatives:

(32) a. Mary is physically stronger than John.
    b. This building is structurally weaker than that one.

Interestingly, as noted by Embick, aspect adverbs are also the only class of adverbs that can appear postadjectivally, as in (33c), which is the word order expected if the synthetic forms are derived by head movement. As Embick also notes, manner adverbs can appear postadjectivally in the positive form as well, as (34) illustrates:

(33) a. *Mary is the smartest amazingly person in the class.   Degree
    b. *Jackie is smarter unobtrusively than Rose.   Manner
    c. This building is weaker structurally than that one.   Aspect

(34) Mary is strong physically.

Embick 2007a further hypothesizes that the optional postadjectival position is somehow connected to the ability of an adverb to outscope the comparative morpheme. Sharpening this observation, I propose that the postadjectival position of an adverb results from adjunction to DegP. Independent support for the ability of aspect adverbials to attach to DegP comes from the fact that with analytic comparatives and superlatives, where no sort of movement is assumed to have taken place, aspect adverbs can appear either above or below *more*:

(35) They are both technically proficient guitarists, but . . .
    a. Cindy is a more technically proficient guitarist than Rick.
    b. #Rick is a technically more proficient guitarist than Cindy.

Unsurprisingly, a difference in order entails a difference in interpretation: (35a) attributes to Cindy a higher degree of technical proficiency, while (35b) claims that Rick is more proficient from the technical standpoint only, though the distinctions become more subtle with different adverbs. PP correlates of aspect adverbs behave similarly: only the analytic form is compatible with the PP modifier treated as part of a complex concept:

(36) a. The verb *go* is more light with respect to its phonology than the verb *fall*.
    b. The verb *go* is lighter with respect to its phonology than the verb *fall*.

(37) a. This book is easier in a hard-to-define way.
    b. This book is more easy in a hard-to-define way.

Here also, the analytic form is norm-related, though the effect is obscured by the fact that the positive form in contradictory statements like (34) can be interpreted stereotypically (see also note 18):

(38) a. I'm more technically proficient than I used to be, but I'm not
          $^{??}$(really) proficient.
      b. Though neither kid is $^{??}$(really/truly) physically strong, Anna is more
          physically strong/#physically stronger than Liz.

I conclude that, like manner adverbials, aspect adverbials modify the positive form of the adjective, adjoining to FP in Rett's model (39a) or to a nonscalar AP, whose head cannot move to Deg°, in Krasikova's model. An aspect
adverbial adjoining to the comparative DegP, as in (39b), does not block synthetic comparative formation:

(39) a.

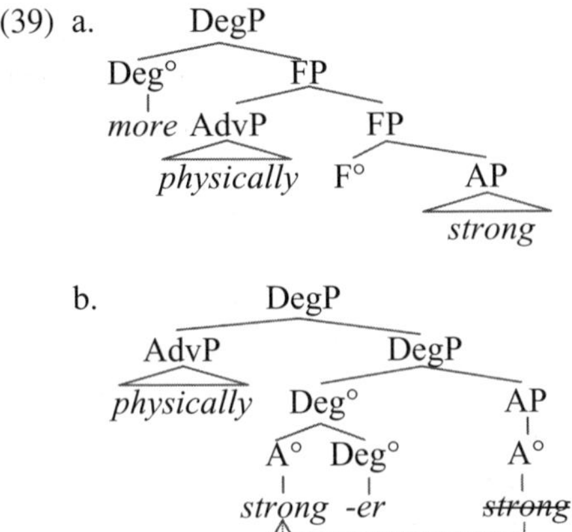

      b.

To summarize, in this section I argued that the derivation of synthetic comparatives and superlatives is constrained by the semantic properties of the AP.
In particular, I showed that in English intersective nonscalar adjectives, such
as *French*, cannot give rise to synthetic forms. I attributed to the same scalarity
constraint the fact that analytic comparatives and superlatives of scalar adjectives are norm-related and then demonstrated that comparatives of APs modified by adverbs or PPs also are.

4.4 Conclusion and Further Questions

In this chapter I provided evidence in favor of deriving synthetic comparatives
and superlatives by the syntactic process of head movement, as opposed to a
postsyntactic process, such as Local Dislocation or Morphological Merger. As
I showed, the former cannot account for comparative suppletion (Bobaljik
2012), while the latter cannot easily deal with phonological constraints on the
synthetic forms. Jackendoff's coordinated comparative construction cannot be
derived by any lowering operation, which also rules out the structural alternative with the comparative DegP in [Spec, AP]. Conversely, the copy nature of

head movement can be used to account for phonological constraints on adjectival stems (contra Embick and Noyer 1999, 2001).

I have also argued that the blocking effect of adverbial modification is semantic in nature: empirically, modified APs are norm-related. Assuming that norm-related APs are nonscalar allows us to assimilate modified APs to nonscalar APs, which do not allow synthetic forms in English. The existence of a semantic constraint on the formation of synthetic comparatives and superlatives is also more compatible with a syntactic rather than a postsyntactic (PF) derivation on the assumption that purely semantic features, such as [degree], are not available at PF.

There are some indications, however, that nonscalar adjectives and norm-related APs do not always pattern the same. Thus in German and Dutch nonscalar adjectives give rise to synthetic forms, while comparatives and superlatives of norm-related APs either can only be analytic (Dutch) or are ineffable (German). It is tempting to hypothesize that the difference between Dutch and German is due to the general availability of analytic comparatives in the former (a relatively recent innovation), but more research is required in order to determine whether languages with both synthetic and analytic comparatives and superlatives available always pattern the same in disallowing the former option for modified APs.

Another issue that I have not discussed here is that of double comparatives, as in *the most unkindest cut of all* (Corver 2005; González-Díaz 2007), which can in principle result from Affix Hopping followed by *much*-support, but do not seem compatible with a head-movement analysis. Linked to that are multidependent comparatives like *John is (much) taller than Mary than Bill is*, which Bhatt and Pancheva (2004) analyze as involving haplology of a double synthetic comparative, though their acceptance by native speakers is varied. I leave these matters as a topic for further research.

Notes

The first instantiation of this paper, a reaction to Embick and Noyer 1999, was presented at GLOW XXIV (April 8–10, 2001, Braga, Portugal). The paper benefited from many insightful comments from Rajesh Bhatt, Noam Chomsky, Kai von Fintel, Ken Hale, Morris Halle, Sabine Iatridou, Tania Ionin, David Pesetsky, Carson Schütze, Marcus Smith and Dominique Sportiche, as well as from the GLOW audience. After a 10-year gap, I returned to the issue of synthetic comparatives and superlatives, using their morphosyntactic properties to argue against the prevalent semantic take on the internal structure of comparatives placing the degree morpheme in [Spec, AP]. As a result, this article has profited from questions and critiques at the Leiden SyntaxLab (March 17, 2011), the Degree Workshop at Sinn und Bedeutung 16 (Utrecht, September 5, 2011), and the Vagueness Circle (Amsterdam, October 7, 2011). I'm also extremely grateful to Eddy Ruys for his ever-available help and advice, as well as to Jonathan Bobaljik and an anonymous reviewer for a most valuable discussion.

The shortcomings of this article in no way reflect upon my enormous intellectual debt to Morris Halle and Alec Marantz for teaching me how to use the wonderful tool of morphology.

My research was generously supported by NWO (Netherlands Organization for Scientific Research, project number 276-70-013).

1. The third approach, which I will not discuss here for reasons of space, is to derive synthetic forms in the lexicon (Poser 1992). The same result can be achieved in the narrow syntax by head adjunction at Merge. Space limitations prevent me from discussing the matter further.

2. This well-known constraint is often violated: quite a few disyllabic adjectives form synthetic comparatives but do not have a light second syllable (see section 4.2 for some discussion). Conversely, Kytö and Romaine 1997 and Hilpert 2008 show that comparatives and superlatives of trisyllabic adjectives found in the British National Corpus are necessarily analytic, though a handful of exceptions, especially for superlatives, can be found.

3. Local Dislocation is sensitive to structure, as well as to the phonological form, which makes it possible for Embick to account for the fact that in metalinguistic comparatives, such as (i) from Bresnan 1973, only analytic forms are possible.

*(i) I am more angry/*angrier than sad.*

Bresnan 1973 attributes the obligatory *much*-support in (i) to a structural difference from the more standard degree constructions and in particular, to nonadjacency. A similar analysis is adopted in Embick 2007a, which proposes that the comparative combines with a null adverb rather than with the adjective. Both structures are incompatible with Local Dislocation, head movement, or Morphological Merger, yielding the impossibility of synthetic forms in all three approaches.

4. Following Sproat 1985, Embick 2007a treats linearization as a two-step process. The first step of fixing local linear precedence relations is followed by the second step: a concatenation procedure, whose result serves as input to Local Dislocation. This refinement does not affect the argument here.

5. Following the standard conventions, I assume that a category-free root projects as a sister of a categorizing x head. Following Kennedy and Svenonius 2006, I hypothesize that the degree argument of the adjective is introduced by *a*, which therefore comes in at least two flavors: the scalar and the nonscalar ones. For the sake of simplicity, the thematic subject of the adjective is not indicated; where the complex morphological nature of the adjective is irrelevant, I will use the standard A and AP notation.

6. If the visibility of the affixal status of a given functional head in syntax is a violation of strict modularity, the alternative is to derive the affixal status of a head as a consequence of its ability to trigger head movement.

7. As noted by Poser 1992, Basque verbal morphosyntax exhibits a synthetic/analytic imperfective paradigm that gives rise to exactly the same sort of issues: synthetic forms of the present and past tenses are only available for a handful of verbs while the rest must use periphrastic forms. Crucially, as discussed by Arregi 2000, there is no

systematic semantic distinction between the two classes of verbs and for both the periphrastic forms can be used to express habituality. Arregi argues, contra Laka 1993, that the synthetic forms are derived by V-to-T movement, which is therefore sensitive to the choice of particular lexical items, however this sensitivity is achieved. The same point can be made in relation to the ability of the lexical verbs *be* and (in some dialects of English) *have* to undergo head movement. Because the focus of this chapter is on lowering and/or postsyntactic operations rather than Late Insertion, I will not pursue the matter here.

8. The structure in (7) has been mostly used for comparatives, with evidence drawn from ellipsis resolution and scope interaction with intensional verbs; for the (in)applicability of the same motivation for superlatives, see among others Stateva 2000a, 2000b, 2002, 2003; Sharvit and Stateva 2002; Matushansky 2008.

9. Jackendoff's examples are compatible with the structure in (7) if the analytic (8a) and the synthetic (8b) forms have the different underlying structures in (ia) and (ib), respectively. However, if both structures in (i) are available, examples like (ii) are incorrectly predicted to be grammatical (with the same interpretation):

(i) a. *[AP [ConjP [DegP more] and [DegP more]] beautiful]*
     b. *[ConjP [AP [DegP -er [AP pretty+er]]] and [AP [DegP -er [AP pretty+er]]]]*
(ii) a. **[AP [ConjP [DegP more] and [DegP -er]] pretty +er] = (ia) with Local Dislocation*
     b. *#[AP [ConjP [DegP more] and [DegP more]] pretty]*
                          = (ia) with affixation failure and much-support
     c. *#[ConjP [AP [DegP more [AP beautiful]]] and [AP [DegP more [AP beautiful]]]] = (ib)*

Furthermore, though Jespersen 1956 claims that a conjunction of a synthetic and an analytic comparative, as in (iii), is possible, Mondorf 2007 demonstrates its extreme rarity across different corpora, in the synchronic as well as diachronic perspective. I hypothesize that in modern-day English a conjunction of two comparatives with the same lexical head (and therefore the structure in (ib)) is impossible:

(iii)   *grow bolder and still more bold*

To the extent that (iii) is available, its interpretation is different from Jackendoff's coordinated comparative, suggesting that the positive form of the adjective (*bold*) serves as the basis for the second conjunct (see section 4.3 for discussion).

10. The analogous construction in French (i) is clearly syntactically formed and completely incompatible with an across-the-board movement analysis, since it does not involve coordination:

(i) a. *de plus en plus belle*
         from more in more beautiful
     b. *de meilleur en meilleur*
         from better in better

Further investigation of crosslinguistic availability and realization of Jackendoff's comparatives is required to verify the hypothesis that the first conjunct is not itself a comparative morpheme. Importantly, coordinated comparatives share their inability to combine with an overt standard of comparison with the semantically similar comparatives of incremental change (Beck 2000; Zwarts, Hendriks, and De Hoop 2005):

(ii) a. Each subsequent apple was more succulent.   *Beck 2000*
　　b. The final exams get easier each year.   *Zwarts, Hendriks, and De Hoop 2005*
　　c. Wolves get bigger as you go north from here.   *Carlson 1977*

I leave this topic for future research.

11. Marantz 1988 addresses the bracketing paradox in (12) in the context of Morphological Merger. However, because Morphological Merger is not conditioned by linear adjacency, the prefix *un-* doesn't function as an intervener. Bobaljik 2012 proposes that the prefix *un-* is transparent with respect to the diacritic feature [+m].

12. The A-to-A prefixes *over-*, *extra-*, and *super-* are also incompatible with synthetic forms, but here not only their prosody (they are all disyllabic and accented), but also their semantics (see section 4.3 for some discussion) might be responsible. The monosyllabic English prefixes *re-* and *de-* do not form adjectives. Others, such as *anti-* or *pre-*, do not combine with adjectival stems, while the negative *in-* selects for Latinate stems, which are derived and therefore tend to be at least disyllabic.

13. The remaining disyllabic adjectives with word-final stress in Hilbert's list are *absurd* (16/1), *compact* (18/4), *corrupt* (5/1), *intense* (169/4), *mature* (141/14), *obscure* (40/2), *polite* (23/7), *robust* (95/1), *secure* (156/4), *severe* (227/9), and *sincere* (12/2), with the numbers in parentheses indicating the word counts of analytic and synthetic comparatives in the British National Corpus. Synthetic superlatives of longer adjectives are noticeably more frequent than their synthetic comparatives, though their use is also clearly emphatic in some way, approaching that of absolute superlatives, aka elatives. Whatever additional factors facilitate the formation of synthetic superlatives with adjectives that resist synthetic comparatives, the pragmatic effect accompanying it is difficult to reconcile with a postsyntactic derivation of synthetic forms.

14. While Newell's (2005) Late Adjunction analyses of bracketing paradoxes necessitates a different structure (with Neg° adjoined to a°), Embick's (2007a) hypothesis that Vocabulary Insertion at Neg° takes place after Local Dislocation extends the notion of postcyclic operations to post-spell-out syntax. Because of space limitations, I will not investigate the matter any further.

15. Empirically, the acceptability of AP-internal superlatives appears to depend on the possibility of retaining the superlative interpretation in the absence of the associated definite article—that is, in the predicate position and the DP-internally, as witnessed by the restricted distribution of *[(*the) best]-known* versus the less constrained *(the) [most well-known]*. Due to space limitations, I set this matter aside.

16. The property of norm-relatedness (Bierwisch 1989; Krasikova 2009, 2010) is also known as "comparative presupposition" (Kiefer 1978), orientedness (Seuren 1984), or evaluativity (Doetjes, Neeleman, and Van de Koot 1998; Rett 2008). A degree construction can be interpreted in a norm-related way for a number of reasons. Thus in English norm-relatedness is a property of equatives and interrogatives of negative adjectives (Bierwisch 1989; Rett 2008), as well as of certain cases of cross-polar nomalies (Bierwisch 1989), while in Russian norm-relatedness obtains regardless of the polarity of the predicate (Krasikova 2009, 2010). We set these issues aside here.

17. Norm-relatedness is not the only reason for the appearance of the analytic form with short adjectives. For other factors inducing its use, such as metalinguistic interpretation (Bresnan 1973), prosody, and so on, see Kytö and Romaine 1997, Lindquist 2000, Mondorf 2002, 2003, 2009, and Hilpert 2008. Assuming that the scalar and the norm-related readings of an adjective are not in any sort of competition derives the optionality of analytic forms in contexts where the positive is true.

18. More complex are examples like (i). On the one hand, intuitively the use of low-degree adverbs like *somewhat* or *slightly* means that the positive AP cannot be used. However, an attempt to explicitly negate the positive form leads to a perceived contradiction, unless the second instance of the adjective is interpreted in an emphatic way, as if it had been modified by the adverb *really*. I leave the matter for future research.

(i) a. #Isabelle is slightly drunk, but she's not drunk.
    b. #The issue is somewhat unusual, but it is not unusual.

19. Here also, comparatives formed with the freestanding degree morpheme *more* give rise to an alternative bracketing, where *more* combines with the adverb to the exclusion of the adjective, as well as to a metalinguistic interpretation (where *more* can be replaced by *rather*). Both these readings will be disregarded here.

5 Is Word Structure Relevant for Stress Assignment?

Tatjana Marvin

5.1 Introduction

In this chapter, I compare two different positions as to the relevance of structure when it comes to stress assignment in English derived words: the "classic derivational" and the optimality theory (OT hereafter) approach. The central issue is how to account for the preservation of stress (and vowel quality) in English affixation and whether the structure of derived words plays any role in the process. As is well known, words in English may but need not change the position of primary stress when affixed. In (1) the word *government* is derived from the word *govern* and the position of primary stress is not changed in derivation. When *governmental* is derived from *government*, primary stress shifts from the first to the prefinal syllable. In (2), primary stress is shifted in each of the two derived words, *instrumental* and *instrumentality*.

(1) góvern   góvernment   gòvernméntal
(2) ínstrument   ìnstruméntal   ìnstrumèntálity

Furthermore, even if the position of primary stress is changed with affixation, the syllable bearing primary stress at some point in the derivation preserves the stress in the form of secondary stress (e.g., gòvernméntal, ìnstruméntal) and is at the same time prevented from being reduced to a schwa (e.g., ìnstrumèntálity).

In this chapter, I first look at an analysis of the phenomenon as in (1) and (2) that can be seen as a natural extension of the classic derivational analysis found in Chomsky and Halle 1968 (SPE hereafter). The latter is supplemented with a theory of word structure as in Marantz 2001 and will be referred to as a phase analysis of the preservation of stress. The proposal together with its theoretical background is presented in sections 5.2–5.4. In section 5.5, the chapter presents a conceptually very different analysis of the phenomenon, the OT analysis of English word stress in Burzio 1994, comparing (in section 5.6) the two proposals in the relevant points. Since stress assignment in English is too vast a topic to be covered here, this chapter only aims at commenting on the role of word structure in both approaches in general terms, and presenting the difference in the approaches on a very limited set of data. Also, the approaches are summarized (relatively) in detail only with respect to stress-structure connection, while only crucial points are highlighted otherwise.

5.2 Phases at the Word Level

The first view presented in this chapter is couched in the framework of Distributed Morphology (Halle and Marantz 1993; Halle 1997a; Marantz 1997). In this approach, words are built by the syntax performing all merger operations, while roots and affixes have no category per se, but are merged in the syntax with category-forming functional heads such as the "little" n, v, a, to form nouns, verbs, and adjectives, respectively. For example, the verb *read* is constructed by taking the category-neutral root $\sqrt{\text{READ}}$ and attaching a phonologically null little v; the derivation can then proceed by further attachment of a little a realized by *-able*, giving us the adjective *readable*, and so on.

If words are treated as a result of syntactic processes, they are "expected" to show similar syntactic phenomena as found on the sentence level. One example is Chomsky's (2001) phase-by-phase spell-out, where the main idea is that derivation of sentences proceeds in phases, where phases can be seen as (predetermined) chunks of syntactic structure that are spelled-out at the phonological and semantic level and are after that point impenetrable for potential semantic or phonological operations that the structure built on top of them might cause. Marantz (2001a) proposes that phrases such as little vP, nP, and aP constitute such chunks—that is, phases at the word level. The idea (elaborated in Marvin 2002) is summarized in (3):

(3) a. Phrases headed by word-forming functional heads, such as little v, little n, and little a, constitute spell-out domains at the word level (Marantz 2001a).

  b. Phases at the word level are subject to Chomsky's (2001) Phase Impenetrability Condition.

  c. *Phase Impenetrability Condition at the word level* (PIC hereafter)*:* H and its edge (specifiers, adjoined elements) are spelled out at the next strong phase. The domain of H is spelled out at the phase of HP. A head h adjoined to H is in the domain of H.

At the point of the merger of the category forming head x (where x stands for v, n, or a, and is supplied with derivational affixes), the complement of the little x is spelled out (i.e., meaning and pronunciation are determined) and from that point on it is inaccessible to heads attaching higher. The mechanism is schematized in (4).[1]

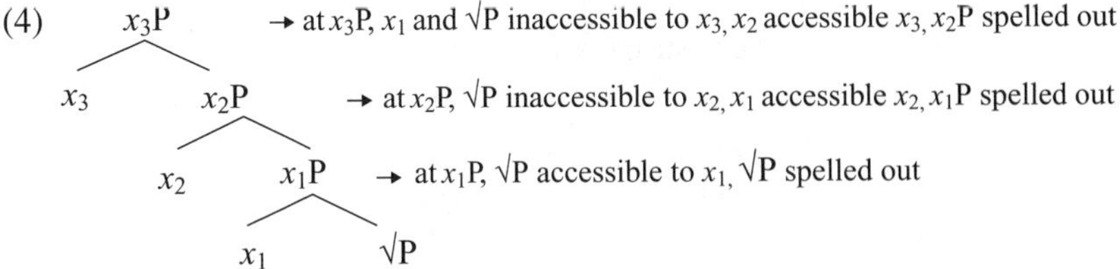

5.3   Stress Assignment in English

5.3.1   English Stress in General

In this section I present how stress assignment in English is captured in a rule-based approach; for a phase analysis (in section 5.4) the analysis adopted is as in Halle 1998.[2] Relying on the metrical theory in Idsardi 1992 and Halle and Idsardi 1995, Halle 1998 proposes that the English stress system is constituted by the Main Stress Rule (MSR) supplemented by two edge-marking rules.[3] The MSR has two parts. A binary foot is constructed at the end of a string whose last asterisk projects a light syllable. A unary foot is built if the last syllable is heavy or there are not enough syllables in the word to construct a binary foot. The application of the MSR is illustrated in (5). For example, in the word *develop*, the last syllable is light and so a binary foot is constructed: *de(velop*.[4] In the words *robust* and *cajole*, on the other hand, the last syllables are heavy; therefore a unary foot is constructed (e.g., *ro(bust, ca(jole*). In the rightmost column we find words with only one syllable, where only unary feet can be constructed regardless of the syllable weight such as *(put, (black*.

(5)   *(**   * (*   * (*   (*
      devélop   robúst   cajóle   pút

In addition to being subject to the MSR, some words are also subject to edge-marking rules (henceforth EMR), which apply to a list of words before the MSR. The first of the two edge-marking rules, referred to as the RLR EMR, inserts a right parenthesis before the final syllable of the word if the syllable contains a short vowel (6a). The second edge-marking rule, the LLR EMR, inserts a left parenthesis to the left of the rightmost syllable (6b).[5]

(6) a.   *(* *]*   *(*]*   * (*]*   (*]*
          América   agénda   Tacóma   vílla
     b.   (* * [*   * (* [*   * * (* [*   (* [*
          málachìte   stalágmìte   monophýsìte   Hússìte

The EMR rules apply first to a list of words; they are followed by the MSR, which applies to all words. Halle (1998) also assumes that the feet constructed on line 0 are left-headed.[6]

Finally, in Halle's system suffixes are either cyclic (i.e., they trigger the application of stress rules in the constituent they form) or noncyclic (i.e., they do not affect the stress of the constituent they form when attached). This is also adopted by the phase analysis.

5.3.2 The Relevance of Structure for Stress Assignment: SPE

The standard example from SPE illustrating the phenomenon of the preservation of stress and vowel quality in English affixation is the 'minimal pair' *condensation–compensation*. SPE observes that in some dialects of English the boldfaced /e/ in *condensation* reduces to a schwa, while this is not the case in the word *compensation*, despite the fact that phonotactically and morphologically the two words are very similar. The vowel reduction is a consequence of the Vowel Reduction Rule, which reduces a lax vowel to a schwa in English. The explanation for this fact offered in SPE is that the difference between the two nominalizations follows from the stress of their constituents. That is, the nominalizations "contain" the verbs *condense* and *compensate*, which represent the first cycle in the formation of the nouns *condensation* and *compensation* and that have different stress patterns.[7]

(7) a. condensátion (/e/ is not reduced)   condénse → condènsátion
    b. compensátion (/e/ is reduced)   cómpensate → còmpensátion

In the former verb, primary stress is found on *condénse*, which consequently means that the stressed vowel cannot be reduced to a schwa in the cycle of the verb, where stress is assigned for the first time. In the latter verb, main stress is found on the first syllable of *cómpensate*; therefore, the corresponding /e/ in *compensate* can be reduced to a schwa in the cycle of the verb. If the stress from earlier cycles is preserved, we now have a natural explanation for the different vowel quality in the two nominalizations. In *condensation*, the vowel /e/ has received stress on an earlier cycle (i.e., the verb cycle) and is therefore prevented from being reduced. In *compensation*, the vowel /e/ has been reduced on the earlier cycle and has remained the same in the nominalization.[8]

The insight from SPE can be directly linked to a finer-grained (i.e., phase) syntax at the word level as proposed in Marantz 2001a. In such an analysis the phase spell-out and PIC as in (3) apply to stress in connection with vowel reduction. Stress and vowel quality preservation in the sense of SPE and Kiparsky 1979 are a consequence of the phase spell-out mechanism. To sum up, the stress in English words is a result of the interaction of the three factors:

(8) a. Building blocks with selectional and stress-related specification
   b. Set of phonological rules
   c. Structure of building blocks

The building blocks (affixes) have stress-related specification, which means that they can be stress neutral or can trigger the application of stress rules when attached (noncyclic or cyclic, respectively, as in Halle 1998). The set of phonological rules comprises different rules of English stress and the order in which they apply (Halle 1998). Some phenomena, such as the preservation of stress in derivation, cannot be accounted for only by the properties of building blocks and stress rules, requiring for their explanation also the third element, the structure of the building blocks. Specifically, the phase analysis restricts the system in such a way that when stress has been assigned in a particular chunk (little x in our analysis), neither the properties of building blocks attached to the chunk nor the stress rules can erase this information. Once assigned, stress is preserved.

5.4   Rules and Phases Applied to English Stress

In this section, I join the three elements summarized in (8) and claim the following:

(9) a. Words are composed of little xPs (Marantz 2001a).
   b. MSR and EMR apply at every xP if triggered by a diacritic marking on x (i.e., by cyclic affixes); they also apply at the last xP if not triggered before.
   c. Vowel Reduction Rule takes place at the level of "prosodic word".
   d. A "phase analysis" (phase spell-out and PIC) as in (3) applies to stress in connection with Vowel Reduction.

To illustrate this claim, we take the word *governmentalese* with the structure as in (10) and the step-by-step derivation as in (11a–e). At each phase the corresponding metrical grid is indicated after the illustration of rule application by using a diacritic (e.g., *á*) to mark stressable elements that receive asterisks on line 1. The Vowel Reduction rule (which is noncyclic in nature) takes place after the spell-out of n_2P.

(10)

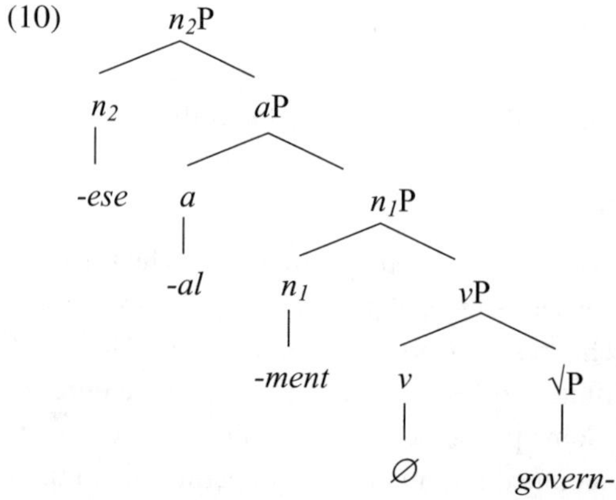

(11) a. At *v*P: *govern-(∅)* → EMR → MSR → the root is spelled out as góvern

    line 1   *
    line 0   (*   ]*
              govern

b. At *n₁*P: *govern(ment)* → the *v*P is spelled out as góvern[9]

    line 1   *
    line 0   (*   *   *
              govern ment

c. At *a*P: *government(al)* → EMR → MSR → the *n₁*P is spelled out as góvernmént

    line 1   *   *
    line 0   (*   *   (*] *
              govern ment al

d. At *n₂*P: *governmental(ese)* → EMR → MSR → the *a*P is spelled out as góvernméntal

    line 1   *   *   *
    line 0   (*   *   (*   * (*
              govern ment al ese

e. at the next higher phase: *governmentalese*: → EMR → MSR → Vowel Reduction → *n₂*P is spelled out as góvernméntalése / gʌvərnmɛntəli:z/

At *v*P, where the root phrase is spelled out, the stress rules apply, which gives the root spell-out *govern*, illustrated in (11a). At the next phase, *n₁*P, the affix *-ment* is added to the structure and consequently to the metrical grid, but since this affix is noncyclic, it does not trigger application of EMR and MSR. Thus the *v*P is spelled out as *govern*, as illustrated in (11b). The next spell-out, illustrated in (11c), occurs at *a*P, where the stress rules are triggered by the attachment of the cyclic affix *-al*. The underlying representation of the affix *-ment* is still accessible at this point, so after that syllable receives an asterisk on line 1, the vowel in *-ment* surfaces as a /ɛ/. The underlying representation of the root is inaccessible at this point, so even if the stress rule would at this point assign no asterisk on line 1 to the root *govern-*, the latter still retains the asterisk due to its already having been spelled out two phases ago, as in (11a). The resulting effect of this mechanism is that it appears as if the stress is preserved from the previous assignments. At n_2P, the complement *a*P is spelled out as *góvernméntal*, as illustrated in (11d). The affix *-ese* attached at this point triggers a reapplication of the stress rules, causing a projection of a line 1 asterisk on the syllable /ese/. However, since the chunk *góvernmént-* is inaccessible to the stress rules applying at n_2P all line 1 asterisks of n_1P are preserved in the grid. Finally, at the next higher phase n_2P is spelled out. Stress rules have assigned a line 1 asterisk to *-ese*, while line 1 asterisks of *a*P are still preserved from previous phases. Since now we reach the end of the word, Vowel Reduction takes place. This is illustrated in (11e).

It should be noted that in this derivation, we do not state explicitly what the difference between primary stress and subsidiary stress is, the term *stress* covering both cases. Obviously, the assignment of primary stress cannot be subject to PIC as in (3), since it can change with affixation. I assume here that primary stress assignment is a phenomenon occurring at lines higher than line 1 in the metrical grid and is therefore not subject word-level PIC. [10]

5.5 English Stress in OT

In this section I present a very brief overview of a non-rule-based approach, with special emphasis on stress assignment in derived words. The approach considered here is an OT-based analysis found in Burzio 1994, which makes use of interactive well-formedness (markedness) constraints in the spirit of McCarthy and Prince 1993 and Prince and Smolensky [1993] 2004.[11] To begin, let us summarize Burzio's (1994) analysis for underived words. Burzio proposes a typology of possible feet as in (12), where "H/L" stands for heavy/light syllables and vowels followed by colons are long.

(12) *Typology of possible feet (Burzio 1994, 165)*

	Feet	Nonrightmost	Rightmost	
a.	mo(nòn.ga)héla	(Hσ)	(Hσ)	a(gén.da)
b.	(wìn.ne.pes)sáukee	(σLσ)	(σLσ)	a(mé.ri.ca)
c.	ac(cé.le)ràte	(Lσ)	#(Lσ)	h(ónes)t

The second component of his analysis is the English-specific existence of a special class of syllables, referred to as weak, which may or may not be metrified, as exemplified in (13a–b), where the weak syllables are italicized. In the words on the right-hand side, the final weak syllable is extrametrical, which is why these words appear as exceptional in terms of stress.[12] If not extrametrical, weak syllables form the so-called weak feet, which fail to attract primary stress, as illustrated in (13c). In English, primary stress falls on the rightmost nonweak foot (if more than one foot is involved), while feet in general are head initial.

(13) a. a ris (to cra *cy*)   (ac cu ra)*cy*
     b. ob(ject *tive*)   (ad ject)*ive*
     c. (órtho)(dòxy), (árchi)(tèct*ure*)

The final postulate of the analysis is that all English words end in a vowel, leading to positing final null vowels in words such as *robust* or *develop*, as shown in (14).

(14) a. ro(bús t∅) / (éarnes)t∅
     b. de(vé lo p∅) / (ás te ris)k∅

5.5.1 Constraint Ranking

5.5.1.1 No Affixation

The highest-ranked constraint for nonaffixed words is the so-called Metrical Well-Formedness, which represents a set of possible feet in English that is summarized in (12). Indeterminacies with respect to parsing—for example, whether a certain chunk is parsed as (σLσ) or σ(Lσ)—or to the metrification of weak syllables are resolved by further constraints ranked lower than Metrical Well-Formedness. These constitute the so-called Metrical Alignment constraint and will not be further summarized in this chapter.

5.5.1.2 Affixation

When it comes to the interaction between stress and word formation, Burzio (1994) claims that stress in morphologically complex words maximally preserves the metrification of the constituent parts and that the principle imposing consistent metrical characteristics on morphemes is "embodied" in a special constraint. The constraint regulating the preservation of stem stress under affixation is Metrical Consistency, ranked as in (15).

(15) Metrical Well-Formedness >> Metrical Consistency
     >> Metrical Alignment

To illustrate a simple case of Metrical Consistency at work, let us take the word *manipulation*, which according to Metrical Well-Formedness could be parsed either as *ma(nìpu)(látion)* or *(mànipu)(látion)*. However, the first parsing wins, because it preserves the stem stress of *ma(nípu)(làte)* and as such crucially obeys Metrical Consistency, violated by *(mànipu)(látion)*.

Burzio notes the two familiar patterns of stress preservation under affixation, which he refers to as weak stress preservation and strong stress preservation, exemplified in (16) and (17), respectively. The first is traditionally argued to occur when stress-changing suffixes are attached, while the second occurs with stress-neutral suffixes.

(16) napóleon → na(pòle)ónic∅

(17) pròpagánda → pròpa(gándis)t∅

The "stress-changing" nature of suffixes is a result of the interplay of Metrical Well-Formedness, Metrical Consistency, and their ranking (see also (22)). The difference between what we traditionally term stress-changing and stress-neutral suffixes (i.e., between (16) and (17)) is a reflex of the way the final syllables in the stem can combine with the phonological structure of the suffix itself either to guarantee preservation of stem stress or not. In Burzio's theory suffixes are assumed to have lexical representations that mark off their position in a foot (e.g., -ic∅), a)l, -ity), -atio)n). Stem stresses are preserved if and only if they correspond to independently well-formed feet.

In the word *napoleonic*, which contains a "stress-changing" affix *-ic*, the stem *napoleon* and the affix *-ic* interact in such a way that the constraint of Metrical Well-Formedness requires a remetrification. Consider the derivation of *na(póleo)n* + ic∅). If the parsing of the stem *na(póleo)n* is to be preserved in this newly derived word, we get *na(pòleo)(níc∅)* and the last foot violates Metrical Well-Formedness.[13] Thus a different parsing has to be employed, one with a ternary foot at the end—*na(pòle)(ónic∅)*—the result of which is perceived as a stress shift, since now primary stress falls on a different syllable than in *na(póleo)n*. However, Metrical Consistency makes sure that the winning candidate, though remetrified, is the one that preserves the stem stress from the word *napóleon* and not perhaps one that does not (e.g., **nápoleónic*).

In the word *propagandist*, with a stress-neutral affix *-ist* and the stem *propaganda*, the syllable from the affix (*-(d)ist*) overlaps with the syllable of the stem (*-(d)a*) and the constraint of Metrical Well-Formedness assigns stress as in (17) in a predictable fashion; no remetrification is needed, which results in an apparently complete preservation of stress.

To sum up, in Burzio's system the default case is that stress does not change with affixation (and that morphemes keep their stress throughout the derivation) unless the phonological properties of the stem and affix combination require a remetrification of the whole to obey the higher-ranked well-formedness constraint. Affixes are thus neither stress neutral nor stress changing by nature as in the SPE approach, but appear as such due to the interaction with the stem and the tendency to obey the well-formedness constraint.

5.6 Comparison to Phase Analysis

In this section I compare Burzio's analysis with the phase analysis as proposed in this chapter, with special emphasis on the stress preservation phenomenon. I examine three environments where I believe the phase analysis better captures the facts because of the importance it assigns to structure in relation to stress assignment.[14]

5.6.1 Scope of Stress Preservation

As pointed out by Kager (1995), in Burzio's approach, it is not really clear what the scope of stress preservation within a set of morphologically related forms is. If there is no element ensuring that the phonological shape of a complex word can only depend on that of its embedded morphemes, such as the *cycle* in derivational approaches, then nothing prevents the incorrect prediction that stems preserve stress from their derived forms (e.g., **orígin* from *original*) or that a derived form could preserve stress of any deeply embedded constituent (**òriginálity* if from *órigin* rather than from *original*) (Kager 1995).

The phase analysis does not incur such problems because the 'embedding' relationship between words and their constituent parts is clear from the structure of words. For example, the three words *origin, original,* and *originality* are posited the structures as in (18), from which stress is derived via stress rules "bottom up." The noun *origin* is derived by attaching a phonologically null *n* to the root √ORIGIN. The adjective *original* does (crucially) not embed the noun *origin*, but the root √ORIGIN, to which the little *a* (realized by *-al*) is attached. It is thus not surprising that the root stress is different in the first two words—in both cases the affixation can affect the stress on the root, and it does it in a different way given that in the noun but not in the adjective the affix is null. In *originality*, the stress is assigned first at the spell-out at *a*P (*original*) and then at the spell-out at *n*P (*orìginálity*). Note that at *n*P, the root stress is not changed anymore (it is retained as secondary stress); only the stress in the "not-yet-spelled-out" constituent is affected (the one on *-al-*). In the phase approach, what might be problematic is finding independent evidence for positing a certain structure for a derived word, but not the directionality of stress preservation.[15]

(18) Origin–original–originality

Origin	*Original*	*Originality*

*n*P
 	
  n   √P
 	
	|
	  Ø *origin-*
		*a*P
		
		  a √P
		
			|
			 -al *origin-*
				*n*P
				
				  n *a*P
				  |
				 -ity *a* √P
				
					|
					    -al *origin-*

5.6.2 Stress Preservation and Mixed Suffixes

Let us now turn to the case of the so-called mixed suffixes, such as *-able* or *-ist*. These are suffixes that (according to Burzio's definition) in some words appear as stress changing (e.g., *rémedy–remédiable*), while in others they appear as stress neutral (e.g., *prevént-prevéntable*).[16] Burzio (1994) solves this "puzzle" by referring to the metrical properties of the combination of the stem and the affix, claiming that *-able* is stress neutral if the first syllable of *-able* replaces the null vowel of the stem, as in (19). Otherwise, *-able* appears stress changing, since it requires remetrification as in (20).

(19) a. pre(véntØ) → pre(vénta)ble
     b. in(hábitØ) → in(hábita)ble

(20) a. (démon)(strá:te) → de(mónstra)ble
     b. (ímpreg)(nàte) → im(prégna)ble
     c. (rémedy) → re(mé:dia)ble
     d. (súrvey) → sur(véya)ble

In the phase analysis the mixed behavior of suffixes such as *-able* can be explained as directly related to the structure of the word, as a consequence of the attachment site of the affix. To illustrate the idea, consider a well-known fact that base verbs in *-ate* are often (e.g., *operable, irritable, navigable*), but not always truncated (e.g., *cultivatable, emancipatable, operatable*) (Plag 1999). Apart from having a special stress when compared to their nontruncated counterparts, the truncated forms can also have a special—that is, unpredictable—meaning (*operable* = 'usable'), while the nontruncated forms always have a predictable meaning (*operatable* = 'that can be operated'), following Marantz 2001a. The claim here is that the stress pattern of the nontruncated adjectives is predictable (*òperátable*), which follows from the structure of these words, as in (21b). The affix *-able* is not attached to the root directly, but rather to an already-made verb *óperàte*, or in other words, it is introduced after the first phase (*v*P), when the stress on the root has already been negotiated. Given (3), the stress of the spelled-out constituent is preserved, in this case as a subsidiary stress on the root. The so-called truncated form could in principle have a different stress on the root because of a different structure: here the affix attaches not to an already-made verb, but to the root and can influence the spell-out of the root.

(21) a.

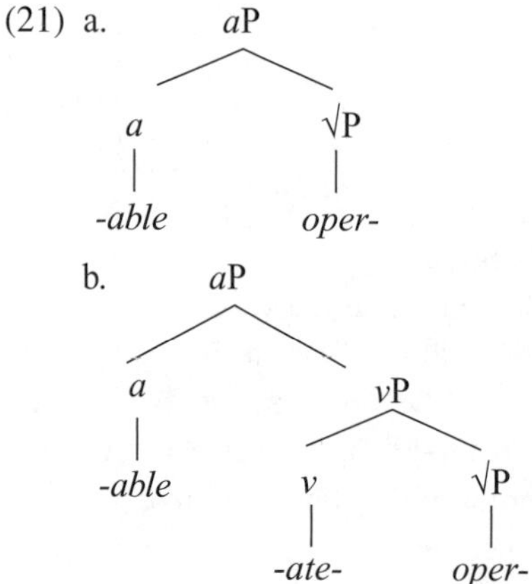

This is perhaps even more obvious in pairs such as *cómparable* 'roughly similar' and *compárable* 'that can be compared', where a special meaning and stress is seen in *cómparable* (structure in (21a), not related to the verb *compáre*) and a predictable meaning and stress is found in *compárable* (structure parallel to (21b), related to the verb *compáre*, with a null *v*).

The same reasoning is employed with respect to Burzio's examples in (20). In the words *demonstrable* and *impregnable* no stress shift occurs with respect to the verbs *demonstrate* and *impregnate*, because the derived adjectives are not derived from the verbs in *-ate*, but rather have the structure as in (21a). Similarly, for *remedial*, I posit the structure as in (21a); this word is not derived from the noun, but from the root √REMEDY. *Surveyable* in (20d) is different; its structure (and meaning 'that can be surveyed') is parallel to (21b), with the verb *survéy* as an intermediate constituent, from which stress is retained.[17] Burzio's examples in (19) have the structure parallel to (21b), where no stress shift is predicted anyway.

Finally, it is interesting that Burzio (1994) also notes the following generalization: mixed suffixes are systematically neutral with free stems but not neutral with bound ones, the property that in his system is captured by positing the two constraints as in (22).

(22) *Metrical consistency hierarchy: a > b*
    a. Stem consistency
    b. Suffix consistency
                                                          Burzio (1994, 254)

Mixed suffixes appear neutral when attached to a free stem, satisfying (22a) (e.g., *americanist* derived from *american*), but impose a fixed stress pattern otherwise—that is, with bound stems (e.g., *antagonist* derived from *antagon-*), satisfying (22b), when (22a) is irrelevant. The observation about the relevance of bound and free stems, which is a structural one, is translated into the two constraints in (22) as if its structural nature were irrelevant. In the phase approach, this observation follows directly from the relevance of word structure—an affix can influence the spell-out of the root if attached to the root directly (bound stem, e.g. (21a)), but not if attached after another *x*P has been attached to the root (free stem, e.g. (21b)).

5.6.3　Stress Preservation in Multiple Suffixation

The next structure-related phenomenon, which is stated just as a coincidental (though important) observation in Burzio 1994, is stress preservation in multiple suffixation. Burzio (1994) observes that neutral suffixes maintain their word integrity under further suffixation–if in the structure (23) suf_1 is neutral, then suf_2 can affect at most suf_1, but not the stem, as seen from example (24).

(23)　stem + suf_1 + suf_2

(24)　desíre → desírable → desìrabílity

In the phase analysis such generalization is directly predicted. Given (3) and (4), attached material can influence the stress of the previous phase, but not of the phase below the previous phase. In (23) it is therefore expected that suf_2 could influence the stress placement in the material in the suf_1, but not in the stem, because the latter will be spelled out at the point where suf_1 is attached and thus become inaccessible for potential changes in further suffixation. What follows directly from the structural position of the affix in the phase theory is coincidental in the OT analysis, yet then stated as a generalization described above.

5.7　Conclusion

So, is word structure relevant for stress assignment in English? To some extent the answer is yes in both of the two analyses. In the phase analysis, structure is one of the crucial factors when it comes to explaining the preservation of stress, especially the behavior of mixed suffixes and the range of influence that a certain suffix can have in a multiply suffixed word. In the OT analysis structure is relevant in an indirect sense—stress in English words is assigned differently in affixed and nonaffixed words (i.e., further constraints are needed for affixed words), so word structure (whatever it may look like) cannot be completely ignored. However, only the phase analysis can predict and account straightforwardly for the mixed behavior of certain affixes or the range of influence affixes can display as following from the structural position that the affix takes in the word. In the OT analysis such instances appear either coincidental or are made to follow from further constraints that indirectly rely on the structure of words.

Notes

For insightful comments and discussion on this topic, I would like to thank the audience at the workshop *What's in a Word* (University of Tromsø, September 2010). Thanks also to the two anonymous reviewers and the editors for providing numerous helpful comments and suggestions.

1. For a similar view, see also Embick 2010. See Lowenstamm 2010 for a critical assessment of views where derivational affixes are categorical exponents and phases within words correspond to SPE cycles, such as Marvin 2002 and Embick 2010.

2. In principle some other theory of stress assignment could be adopted to illustrate the same point (e.g., Halle and Vergnaud 1987, where extrametricality is used instead of edge marking).

3. Here I adopt a theory of stress following Liberman 1975, Prince 1983, Halle and Vergnaud 1987, Idsardi 1992, Halle and Idsardi 1995, and Halle 1997b, in which stress contours of words are expressed by means of a metrical grid. The bottom line (line 0) is composed of projections of the stressable elements (syllables), and higher lines constructed by projecting some of these elements upward (the heads). Feet are constructed by boundaries (left or right parentheses); they are left- or right-headed.

4. The last syllable in *develop* is considered light because it forms the so-called weak cluster, defined in SPE as a string that consists of a simple vocalic nucleus followed by no more than one consonant.

5. The examples in (6b) have two stresses, one primary and one secondary. They are subject to the Rhythm Rule (Halle 1998), which places the main stress on the leftmost syllable on line 1 of the metrical grid. The Rhythm Rule in Halle 1998 is an extended Rhythm Rule proposed by Liberman and Prince 1977, applying in word sequences as well as within single words.

6. A few other minor rules are needed to account for certain groups of exceptions, for which the reader is referred to Halle 1998. Also, it is important to note that the rules proposed by Halle 1998 have lexical exceptions (e.g., (i) and (ii)). These words are marked in the lexicon as special and cause problems to any theory of English stress.

(i) (*] *   *Also: modest, solemn, modern, auburn, covert (unexpectedly*
    góvern   *subject to RLR)*

(ii) (* *] *
    Cáthol-ic   *Also: Arabic, politic (unexpectedly subject to RLR)*

7. The SPE notion "cycle" can be roughly defined as the point of the application of stress rules to a certain constituent.

8. As to the Vowel Reduction Rule, SPE places it after the process of stress assignment within the word (i.e., in the noncyclic block).

9. The square bracket is added to *govern* as part of the EMR rule that applies to this lexical item; see note 6. Since this bracket is only relevant in the first phase, when the root is spelled out, it is not kept afterward as part of the notation. The same applies to the square bracket related to *-al* in (11c).

10. Similarly, the so-called Rhythm Rule (Liberman and Prince 1977; Kiparsky 1979) can change the properties of the already spelled-out constituent, as can be seen in (i) for the word *thirteen*.

(i) a. *thìrtéen*
    b. *thírtèen mén*

11. This section is intended merely as an illustration of the main principles rather than as a detailed summary of the whole of the Burzio 1994 proposal.

12. Burzio's analysis employs extrametricality confined to special syllables in the spirit of Liberman and Prince 1977. In Burzio 1994, the properties of weak syllables are (tentatively) attributed to their acoustic weakness.

13. (Lσ) cannot appear in word-final position.

14. It should also be noted that Burzio (1994) does not see stress preservation linked to vowel reduction in contrasting pairs such *cond*[e]*nsation* and *comp*[ə]*nsation*. He claims that the difference in vowel length follows from constraints of the preservation of segmental quality (from cond[e]nse and comp[ə]nsate) that is independent of preservation of stress. In this chapter I limit the comparison between his analysis and the phase analysis to the position that the two take to the preservation of stress. For a detailed proposal, see Burzio 1994, chaps. 4, 10, as well as Burzio 2007.

15. There is a small group of exceptions that cannot be explained by a phase analysis. If in (i) the right-hand word is derived from the left-hand one, where the latter is already an *x*P, then we run into a problem of derived words either losing or acquiring stress on the parts that should already be spelled out. This problem is acknowledged in SPE, which reflects the view that groups of words like *solidity* are not the general case, but exceptions hard to capture in any theoretical approach.

(i) a. *sólid–solídity*
    b. *télegràph–tèlégraphy*
    c. *catástrophe–càtastróphic*
    d. *compónent–còmponèntiálity*

16. For a discussion of the "pair" *remedy–remediable*, see also Steriade 1999. Her view, termed *Lexical Conservatism* and similar to Burzio's approach in spirit, will not be discussed here due to lack of space.

17. The phasal approach has to account for the difference between the stress of the noun *súrvey* and the stress of the verb *survéy*—here some rule connecting such pairs of nouns and verbs needs to be posited.

6 Locality Domains for Contextual Allomorphy across the Interfaces

Alec Marantz

6.1 In the Beginning Was the Q

The origins of Distributed Morphology can be traced to an argument Morris Halle and I had when I arrived (back) at MIT to teach in the fall of 1990. I came with "lexicalist" assumptions about morphology, as worked out for example in Lieber 1992—not the notion that words were built in the lexicon but rather the notion that lexical items, identified by their phonology, brought syntactic and morphological features with them into the derivation. Morris was working out a proposal that morphemes with suppletive (phonologically unrelated) allomorphs, like the English past-tense morpheme, were "abstract," as he put it, finding their phonology after the syntax had done its work. Such morphemes he proposed to call "Q" in the syntax, taking over the variable for complex symbols used by Chomsky in *Aspects of the Theory of Syntax*. On the other hand, morphemes without suppletive allomorphy like the English progressive *-ing* were "concrete" and came into the syntax with their phonology, here consistent with lexicalist approaches. Although I was not convinced that we should abandon lexicalist assumptions about morphemes and adopt Qs, I was adamant about one point—if some morphemes were abstract, the grammar would be more coherent if *all* morphemes were Qs (To put it differently, if the only difference between abstract and concrete morphemes was in the multiplicity of spell-out, then concrete morphemes should be treated as abstract morphemes that just happen to have a single realization; i.e., if being a Q did not determine behavior in the syntax, Q-ness should not be a syntactic property of morphemes.) Thus "Distributed Morphology" (DM) in its current form, as associated with Halle and Marantz 1993, 1994, was from its start about the interpretation of abstract morphemes—the building blocks of syntax that found their interpretation in form and meaning in the interpretive components of the grammar.[1] Let history show that Morris was right, but that I made him even more right; to paraphrase a folktale famous in linguistic circles, it was Qs all the way down.

In the Halle and Marantz 1993 theory, Vocabulary Insertion (VI) could be contextually determined, but we did not discuss locality conditions on the size of conditioning contexts for VI that might be principled consequences of the architecture of the grammar. Nor did we envision a melding of DM with Chomsky's Economy framework, critically discussed at the end of that paper. In the following years, with the advent of the Minimalist Program (MP), DM found a natural union with Chomsky's approach, and by the turn of the twenty-first century, the linking of cyclic phase-based syntax within the MP with the architecture of DM was being actively explored.

A number of related lines of research on constraints on Vocabulary Insertion sketched out reasonable theories of constraints on contextual allomorphy—that is, choice between Vocabulary Items (VIs = allomorphs) determined solely by context, where the competing allomorphs realize (are the exponents of) the same set of features on a terminal node from the syntax (cf. competing realizations—allomorphs—of the English past tense, all of which spell out just past-tense features on a terminal node). Bobaljik (2000) proposed a mechanism of root-out VI that guaranteed the downward context could see phonological information and specific VIs while the upward context could only refer to grammatical features (where "up" and "down" are relative to the tree-structure representation of a word). Bobaljik's proposal proved influential as a productive working assumption for research in this area. In the most detailed discussion of the relationship between the syntactic and phonological structure of words to date, Embick (2010) proposed additional locality constraints on contextual allomorphy specifically related to the phonological structure of a word, making reference, for example, to phonological adjacency as defined by concatenation of morphemes. While researchers within the DM/MP universe explored different assumptions about the nature of cyclic spell-out, all work assumed that context would be constrained by phases—that is, only material within a spell-out domain defined by phase heads could be visible as context for VI.

Within the MP, some attention has been paid to the issue of whether semantic and phonological interpretation happen at the same time in a derivation, so to speak—that is, whether the spell-out domains for LF and PF are the same. The most natural assumption within MP is that the same elements from the syntax are sent for interpretation at both interfaces at the same point in a derivation (an assumption sometimes associated with the "single-cycle" hypothesis), but empirical considerations have led linguists to argue for mismatches between LF and PF spell-out domains (see, e.g., Marušič 2009 and the discussion in Grohmann 2009).

In addition to questions about the relative timing of semantic and phonological interpretation within the cyclic derivation of a sentence, in the case of semantic interpretation, we can ask whether there is an equivalent to VI on the LF side that introduces semantic values to morphemes in a manner sensitive to context. That is, is there "contextual allosemy" parallel to contextual allomorphy? Surely "abstract" morphemes are as abstract with respect to semantics as they are with respect to phonology; if morphemes in the syntax were semantically concrete, we would identify them by their semantic content in the syntax rather than by their grammatical features. If there is contextual allosemy when the semantic values of abstract morphemes are fixed at the interfaces, would the locality constraints on contextual allosemy parallel those for contextual allomorphy? This chapter supports the default hypotheses here: that the semantic interface, like the phonological interface, does allow for contextual allosemy, within the same spell-out domains as for contextual allomorphy and governed by the same locality constraints. Section 6.2 briefly explains the locality conditions on contextual allomorphy, adapted from Embick 2010. Section 6.3 presents a quick theory of contextual allosemy, covering the choices of meanings that count here, both for roots and for functional morphemes. For this discussion, it proves crucial to contrast contextual allosemy with the sort of "special meanings" associated with idiom formation. Section 6.4 presents three challenges for the proposed locality constraints on contextual allosemy, two from the literature on Japanese and Greek, and a case parallel to these from English. In this section, it is shown that the data from these languages are actually predicted by the theory of sections 6.2 and 6.3 and highlight an extremely strong prediction about apparent nonlocal conditioning of contextual allosemy. Section 6.5 concludes.

6.2   Contextual Allomorphy

To illustrate what is at issue for locality in contextual allomorphy, consider the "irregular" English past tense *taught*; irregular past-tense morphology in English was one of the prime motivations for Halle's DM precursor theory, with the English past tense an exemplary Q. For Vocabulary Insertion, we want the root TEACH to be realized as /tɔ/ in the environment of the past tense. And we want the past tense to be realized as /t/ in the environment of TEACH (given the phonology of the past-tense stem here, the "regular" past-tense realization would be /d/, after a vowel).[2] For a theory like that of Bobaljik 2000, the contextual allomorphy here fits the generally expected pattern, given the structure in (1): VI at the root is up/outwardly sensitive to the past-tense feature on T, while VI at T is down/inwardly sensitive to the actual item inserted at the root node. However, Svenonius (2012a) has recently questioned whether root-out VI is a necessary addition to the locality constraints on VI already demanded by the cyclic architecture of the MP. Nothing in the present chapter hinges on whether Bobaljik or Svenonius is correct.

(1) a. *English past tense; contextual allomorphy across a phase head*
$\sqrt{\text{TEACH}}$ + v(+voice) + Past = taught
   b. *Overt little v head would block contextual allomorphy of the root*
$\sqrt{\text{QUANTUM}}$ + *ize* + Past = quantized, *quintized, *quantumized, etc.

Two issues of locality are crucial to the English examples in (1). First, there's the question of locality within a phase—that is, putting both tense and the root in the same locality domain. If voice and little v (in combination, or one or the other) are a phase head, then it would seem that the verbal root and tense are on different sides of a phase boundary. Shouldn't the root be spelled out in the complement domain of the v+voice phase head, before the Merger of T? Second, there is a question of adjacency—does the root need to be next to the conditioning environment for VI? Doesn't v+voice interfere with adjacency?

I adopt here a particular version of "derivation by phase" in which the merger of a phase head sends its complement domain for spell-out. For this chapter, we will assume that each root is adjoined to the category head that "types" it as a lexical category prior to any stage of the derivation in which either the category head or the root might be phonologically interpreted. That is, a verbal root will be in the same spell-out domain as voice and little v and will not be spelled out in the complement domain of little v.[3] Only the complement structure of a verb will be spelled out when the voice+v phase head is merged, not the verb root. The assumption that the verbal root falls within the same spell-out domain as the v category head holds important implications for the way that head movement in the grammar might interact with spell-out domains within words, as well as for the theory of verbal argument structure (are the complements of a verb the complements of the root, for example). Work on the relationship between word structure and phonology (see Embick and Marantz 2008 and below) indicates that stems that are also words of a particular syntactic category (e.g., *global* (adjective) in *globalize*) must be spelled out phonologically independent of category-changing morphology that might be attached to them (e.g., *-ize* in *globalize*). As a consequence, if words such as *globalization* are created via head movement of typed categories, with say the adjective *global* moving to the v head *-ize* and then the verb *globalize* moving to the n head *-ation*, the embedded complements to the phase heads— here the adjective complement to v and the verbal complement to n—must be spelled out as the result of merging to the higher phase head. That is, in these cases, the typed stems must be within the spell-out domain of the affix as category/phase head. The head movement here, if head movement is the proper mechanism for word formation, does not serve as an escape hatch for these complements to avoid spell-out within the complement domain of the higher category head. If roots join their category heads via head movement, the relationship between the root and the category head must differ in some way from that of a category head and a lower typed stem, as for the *n* head-*ation* and the typed verbal stem *globalize* in our example; the root and category head are in the same spell-out domain while the typed stem and the category head are not. One possibility to explore in this context is that roots are always ("externally") merged to their category heads and never undergo head movement as part of word formation (although a given root might syntactically merge with different category heads). Other possibilities would involve deeper development of the theory of head movement and syntactic word formation.

As for the intervention of voice and *v* between tense and the verb root, our second locality issue for the English past tense, phonologically null heads do not block adjacency between the root and the conditioning environment for stem allomorphy within a spell-out domain. Embick (2010) discusses the invisibility of phonological zeros for adjacency requirements on contextual allomorphy, but the observation that phonological nulls generally do not stand in the way of relations requiring phonological adjacency is widespread in the literature. In (1b) we illustrate the blocking effect of an overt little *v* on the potential contextual allomorphy of the root or the past tense in English. When the little *v* is spelled out as -*ize*, the past tense cannot trigger contextual allomorphy on the root, nor could the root condition an allomorphic choice of tense.

So, it is not the case that a phase head by itself necessarily blocks contextual allomorphy—the voice+*v* complex between a root and tense does not interfere with Tense serving as the context for VI at the root (and vice versa) since all these heads are spelled out at the same time, in the complement domain of C. If, instead of T, the potential context were a phase head, then a *v*(+voice) node between the root and the context would, in a sense, "block" a contextual relationship between the root and the higher phase head. A simple potential example of this blocking would be English productive -*er* nominalizations, as indicated in (2).

(2) root+*v*(+voice)+*n*   work-Ø-(ø)-er

Here the little *n* node, a phase head under the assumptions of, for example, Embick and Marantz 2008, would send its complement domain for spell-out, including only the root+*v*(+voice) complex, and not the little *n* head itself. The verb would need to find its phonological interpretation independent of the context of the *n* head, which is to be spelled out as -*er* in this example. The *n* head itself could not find the root as context because by the time VI occurs at the *n* head, the root has been realized as part of a bigger structure. Here it does not matter whether the intervening heads are phonologically zero, since the constituent is no longer identifiable as a root. (VI at *n* might be sensitive to properties of the highest head in its complement, which "types" the constituent to which it attaches, so perhaps properties of voice and/or little *v*, for instance, could condition VI at *n* in this example.)

The apparent "regularity" of derivational morphology in English that attaches to word stems (stems that have already been "typed" as a *n*, *v*, or *a*—stems that are not bare roots) has been presented as support for this view that a phase head (a category-determining head) blocks contextual allomorphy triggered by another phase head (category-determining head). See the discussion of inner (root attaching) versus outer (lexically typed stem-attaching) derivations in Marantz 1997 and Embick and Marantz 2008.

(3) "inner" -*er* (rotate/rotor) vs. "outer" -*er* (rotate/rotater)
    "inner" -*ity* (curiosity) vs. "outer" -*ness* (gloriousness)

"Outer" morphology may be phonologically adjacent to a root but never serves as context for contextual allomorphy of the root, nor does the root act as context for contextual allomorphy on the "outer" affix.

Although this section has framed contextual allomorphy as a question of the context for VI—thus for suppletive allomorphs of a morpheme—the considerations here should apply also to contextual allomorphy created by morphologically triggered phonological rules ("readjustment rules"); that is, special changes to the pronunciation of a morpheme with a morphological context, where the morphological identity of both undergoer and context is relevant. (See Harley and Tubino Blanco, chapter 7, this volume, for an argument that suppletive root allomorphy, handled by VI, needs to be separated from phonological readjustments to a previous spelled-out root, with readjustment following VI, although both would be subject to the same locality considerations.) Cases of stem changes in English irregular past-tense formation, which are describable in terms of reasonable phonological rules (or motivated by reasonable phonological constraints; see, e.g., the *feel/felt* alternation), would be governed by the same locality constraints as the *teach/taugh(t)* alternation, where any possible phonological motivation is synchronically obscure and the relevant rules or constraints would be limited to a single item. Morphological context for phonological spell-out must be available in the same spell-out domain and phonologically adjacent to the morpheme subject to the special realization, whether via VI or via what are sometimes called "readjustment rules," whatever the formal realization of these rules within the theory of phonology.

6.3   Contextual Allosemy

Given our understanding of the locality constraints on the phonological spell-out of morphemes, we may ask whether the same constraints apply to contextual allosemy, the possible conditioned semantic spell-out of a morpheme at the LF interface. Since the constraints at the phonological interface were argued to follow from the general architecture of the grammar, we should expect parallel constraints to hold for allosemy. Only material in the same spell-out domain should condition allosemy, and this material should be semantically adjacent to the morpheme whose semantic value is being determined, where semantic "adjacency" would parallel phonological concatenation and mean "semantically combines with directly." But what is contextual allosemy? That is, what sort of meaning alternations could be attributed to a choice of semantic interpretation for a morpheme in a given context?

Of concern in this chapter is root allosemy—the choice of particular meaning for a root in a particular context. However, possible contextual determination of the meanings of functional morphemes is currently being explored in the literature within DM/MP frameworks consistent with the assumptions of this chapter; see Marantz 2012 and Wood 2012, both of which rely on work of Schäfer 2008. The kind of issue being explored in the work on functional heads involves the apparent syntactic and morphophonological leanness of structures that exhibit semantic complexity. For example, although the meaning of transitive *open the door* arguably involves "cause," "change," and "be (in the open state)," the syntax and morphology of lexical causatives like English *open* crosslinguistically fail to show evidence for the full multiplicity of heads implicated by the meaning. In particular, languages do not seem to realize either syntactically or morphophonologically both a "cause" head and an "inchoative" head in lexical causatives, although arguably the interpretation of the little v of an inchoative like *the door opened* would involve "become," with the transitive *open the door*, then, involving a "cause" and "become" meaning. If we limit the basic functional ingredients of the vP to voice, a little v, and the verbal root (where any extra little v would yield a syntactic causative construction and an additional phase), we would need semantic "flavors" of little v to yield different vP meanings (or some way of deriving more complex semantic structures from a limited semantic inventory of heads, given the syntactic context of these heads—the richness of the various contextually determined event interpretations within the vP constituting contextual allosemy on such an approach). Such semantic flavors of v might be contextually determined at the LF interface, rather than being featurally specified in the syntax. Wood (2012) explores a similar approach to the meanings of voice (e.g., the distinction between voice that adds an external argument and voice that does not), extending insights of Schäfer 2008.

For the kind of contextual allosemy of functional heads described above, I know of no challenges to the claim that such allosemy might be governed by the locality conditions suggested at the outset of this section: context within the same spell-out domain as the head being interpreted and semantically adjacent to this head. For example, a little v taking a little vP complement would always be interpreted as a syntactic causative head; such a little v would send its (vP) complement off for interpretation, and for context at its own semantic realization, it would see only the interpreted vP. On the other hand, voice, little v, and the root could interact with an interpreted complement of v to determine their semantic values, with the actual interactions governed by the specifics of how they combine semantically.[4]

Putting aside, then, the extremely interesting topic of functional head allosemy, we turn to roots. In previous work (Marantz 2001), I had claimed that the meaning of a root would be fixed by the first category node up in the tree from the root, the syntactic category node that "types" the root as a lexical category noun, verb, or adjective. There is a sense in which this locality condition follows directly from semantic adjacency—if the root combines semantically with the category head, then this would be the local domain for determination of root meaning independent of phasehood. The types of examples considered in the literature, both in Marantz 2001 and the work to follow, however, also involved phase boundaries as the potential context for meaning determination. These configurations attached an additional, category-changing, head to the category head–plus–root combination, and asked whether the higher head could condition meaning choice of the root over the intervening head. If category heads are phase heads, the higher category head would send its complement for interpretation, preventing it from serving as the context for any contextual allosemy on the root, which spells out in the context of the lower category head. Unexplored were configurations like that in (1) where the root and the category head are joined with a non–phase head, here tense. If these terminal nodes are all sent for spell-out at the same time, could the non–phase head condition contextual allosemy of the root over the category (phase) head?

As an example of the kind of data that supported the phase-based locality constraints on contextual allosemy, consider the verb *to hou[z]e* from the root HOUSE, which shows contextual allomorphy (special voicing of the final fricative in the environment of the little v head) and contextual allosemy (no literal *house* nor even a literal *container* is implied by the verb).[5] One can also make a verb from the noun *house*, with a meaning "do something with houses," as in "He took a bunch of plastic models and housed the room in revenge [filled the room with the houses]." This verb has the "literal" (nominal) reading of the root HOUSE as well as the default phonological form, with a voiceless fricative /s/. Here, the little n head blocks both contextual allomorphy and contextual allosemy because the v head sends the n head plus root for interpretation, without being part of the spell-out domain and thus without providing context for the interpretation of the root. In addition, the root does combine semantically with the little n head, creating the nominal "house" reading. So the lack of contextual allosemy here is overly determined.

The claim that outer derivational morphology cannot determine special meanings for a root has been challenged in the literature. One set of challenges is the topic of the next section; these would constitute true counterexamples if they did in fact involve stacked category-determining heads. However, another set of examples from Borer, Goldberg, Harley, and others (see, e.g., Harley 2011) do not seem to actually hit the mark. Understanding why requires better explication of what root allosemy comes to.

First, we must clarify the distinction between homophony and polysemy. Two different roots, in the linguistically relevant sense, can share a phonological form, yielding homophony as in *bank* (the side of the river) and *bank* (where you keep your money). Much psycho- and neurolinguistic research has contrasted the processing of homophones like *bank* and roots like *table*, which show a variety of related meanings (furniture, organization of numbers and figures, etc.). This work supports the linguistic conclusion that homophones involve separate lexical entries (stored information about the root as morpheme) that just happen to include identical phonological representations, while polysemes involve different semantic interpretations associated with a single lexical entry for a root.[6] While not underestimating the difficulty of deciding whether a given case of ambiguity of interpretation of a root (identified by its phonology) involves homophony or polysemy, nor the problem of understanding how children learn about polysemy, for present purposes we will make a cut between different roots that happen to be pronounced the same (homophony) and a single root, with potentially varied (but related) semantic interpretations (polysemy). The theory of contextual allosemy is the theory about what governs the choice of meaning for a polysemous root.

For the theory of root meanings, it is useful to distinguish at least two dimensions of variation in meaning. First, as explained, for example, by Levinson 2010 among others, roots can be seen as belonging to semantic types associated with the meanings of the category heads n, a, and v. If the usual interpretation of v is to introduce an event variable, of a to introduce a state variable, and of n to introduce an entity variable, then "verbal" roots are those that modify events, "adjectival" roots those that modify states, and "nominal" roots those that modify entities. There are certain natural, systematic meaning alternations associated with these categories such that an entity modifier, for example, naming the product of a verb of creation (*braid*), also has a related use as a "manner" adverbial, modifying an activity (as in "They braided their way through two movies and countless commercials," where the root provides a manner reading; see Levinson 2010). In the context of a little *v* head, then, a root that might normally be given an entity-modifying meaning might receive an event-modifying manner reading, in a sort of type-shifting allosemy. It is of course a matter of intense research and discussion how best to treat this sort of allosemy formally—for example, whether to employ a typed meaning analysis for roots and attribute this allosemy to actual type shifting or whether to employ a different sort of semantic theory that might leave the semantic category of roots unchanged in this type of allosemy. Whatever the ultimate story here, the sorts of meaning shifts for roots identified with this kind of lexical category shift should fall under the theory of contextual allosemy being developed in this chapter.

In addition to potential type-shifting polysemy, there are systematic semantic relations between meanings associated with polysemy that involve related meanings of the same type. So, for example, consider the "abstract" and "concrete" meanings associated with *globe*—meaning a sphere or something spherelike, as well as "the world." In the context of a little *n* head ("a globe"), the root can mean "spherical" (a glass globe Christmas ornament) or "this (whole) planet" (this troubled globe)—no choice is made at this head, so either meaning is possible. If we create an adjective out of the root, *global*, the world/planet reading is picked by little *a*. This reading and not the "sphere" reading then is preserved in further derivation, as in *globalize*, which does not mean "make into a sphere."[7] Similar types of polysemy might be those described as "metonymy," as for *novel*, meaning the artistic work or the physical volume. While either meaning is available for the root *novel* in the context of little *n*, when a verb is formed from the root, the artistic-work meaning is chosen, such that *novelize* means make into a (fictional) story, not make into a physical book.

The claim that the meaning of a root is fixed in combination with a phase (category) head, then, should be understood as the claim that polysemy is resolved in the domain of the first category head combining with the root. Where more than one meaning of a polysemous root is allowed for a given category head, this is an example of "free variation" for semantic combination—the ambiguity appears to be carried up the structure because either semantic variant of the root is possible in the given environment. What is ruled out is semantic flip-flopping—the choice of one meaning in the context of the first category head, with a switch back to a different meaning at the next category head, as in the example of *globalize* switching back to the "concrete-sphere" meaning, after the adjective chooses the abstract "world" meaning in the construction of *global*.[8]

To further clarify the scope of this predicted ban on flip-flopping, consider the nominal, *novelization*. The root *novel*, like other roots that determine reference to the result of writing, can mean either a physical object ("the novel on my table weighs two pounds") or a work of art ("the novel on my table is gripping"). The verb *novelize* seems to resist any meaning that involves the physical-object alloseme of the root; "The magician novelized the hapless author, turning him into a two-pound volume on the bookshelf of his office," requires a very self-conscious, joke reading that is hard to compute. The prediction of the proposed analysis of contextual allosemy is that further derivation cannot choose the physical-object alloseme of *novel* over the verbalizing head that has chosen the abstract reading: *novelization* cannot mean "the creating of a physical book," as in "The novelization of those particular 200 sheets of paper would serve no good purpose."

Apparent counterexamples in the literature to this claim of the local determination of root meanings conflate "idiomatic" special meanings with polysemy. For example, Harley (2011) points out that *nationalize* has a meaning of "make into a government-owned/operated business" that is not predictable from the meaning of *national*. However, the choice of alloseme of the root *nation* in *nationalize* is not distinct from that in *national*; rather, the meaning of *nationalize* is just very specific and tied to world knowledge about governments, and so on. Similarly, to choose an example from Borer, *existentialism*, as a belief system, holds a meaning that is not predictable from *exist* or *existence*, but the allosemes of *exist* and *existence* in *existentialism* seem the same as those in the embedded verb and noun built on the root.

Clearly complex words can acquire special meanings and uses; like phrases, complex words can be idiomatic in the sense of conveying meanings not computable from the meanings of their parts. Marantz (2001) is confusing if not simply wrong in conflating the notion of "idiom" with the notion of "special meaning" or "meaning choice" associated with polysemy. For the issue of root (and likely functional morpheme) polysemy, the relevant locality domain for "fixing" meaning appears to be the phase, while for idioms, the domain is clearly larger. For example, "the bucket," in "kick the bucket," should contain a locality domain in which the root *bucket* finds its meaning among the choices available (here, most likely the physical bucket, as opposed to a "large measure" as in a "bucket of trouble") before the noun phrase is merged into the idiomatic *v*P. Idioms, then, involve a type of meaning that is built on top of polysemy resolution. Harley (2011) and Anagnostopoulou and Samioti (forthcoming) both support Marantz's (1997) claim that idiom formation seems constrained to the domain of an external argument. Since verbs may contain a voice that projects an external argument, derivation on agentive verbs can be idiom-resistant in a way that derivation on verbal roots is not—an important finding, as Harley (2011) makes clear. Here we are interested in cases of derivation that appears to choose a root alloseme over a category-determining head, in violation of the (much stronger) phase-based locality restriction on contextual allosemy.

6.4 Canonical Counterexamples to a Phase Domain for Contextual Allosemy

Although the literature is full of examples of idiomatic "special meanings" spanning phase boundaries in a way that clearly falsifies any equation of "phase" (spell-out domain of a phase head) with the domain of special meanings or meaning determination, the counterexamples to the claim that contextual allosemy is computed within the spell-out domain of a phase head are quite rare. Nevertheless, a particular form of counterexample does appear, in Japanese, Greek, and English. In all these cases, apparent deverbal derivation, built on stems with phonologically overt verbalizing morphology, chooses a meaning for the root that is not built on the meaning of the embedded verb.

Volpe (2005) presents Japanese data precisely as a challenge to the claim that the root meaning must be fixed within the domain of the first phase head up from the root in the derivation. His data are particularly interesting because his analysis builds on an important, well-studied area of Japanese syntax and morphology: lexical causatives and the causative/inchoative alternation. Japanese inchoative/lexical causative pairs built on the same root are famous for showing different patterns of morphological expression. Sometimes it seems that the causative version is morphologically marked with respect to the inchoative version; sometimes the inchoative version is morphologically marked with respect to the causative. And sometimes the causative bears what looks like the regular, syntactic causative ending *-sase*.

Miyagawa (e.g., 1998, 1999) summarizes his and others' work on lexical causatives in Japanese; although aspects of the analysis are open to debate, there seems overwhelming support for analyzing the suffixes signaling either the lexical causative as opposed to the inchoative or the inchoative as opposed to the lexical causative as realizations of a little *v* head attaching to the root. For present purposes, perhaps the most crucial part of the consensus analysis of Japanese lexical causatives is the distinction between lexical causatives, involving a *v* attaching to a root, and syntactic causatives, which arguably involve a higher *v* taking at least a lower *v*P as a complement. There are fairly straightforward ways of identifying lexical causatives and distinguishing them from syntactic causatives. For example, the lexical causative, but not the syntactic causative, may be used in an "adversative" causative construction. And one can create a syntactic causative on top of a lexical causative (by affixing *-sase*), but one cannot create a lexical causative from a lexical causative or a syntactic causative from a syntactic causative (with stacked *-sase*'s). Lexical causatives can have idiomatic readings not predictable from the meanings of their roots, while syntactic causatives always display compositional meanings.

The analysis of lexical causatives consistent with Miyagawa and Volpe involves a little *v* head attaching to a root, where this causative "flavor" of little *v* can be switched for an inchoative "flavor," often spelled out differently phonologically, creating the transitive (causative)/intransitive (inchoative) pairs we see in table 6.1 from Volpe, along with his proposed structure for the nominalizations in the fourth column.

(4)   **n** *d-ashi* 'soup stock'

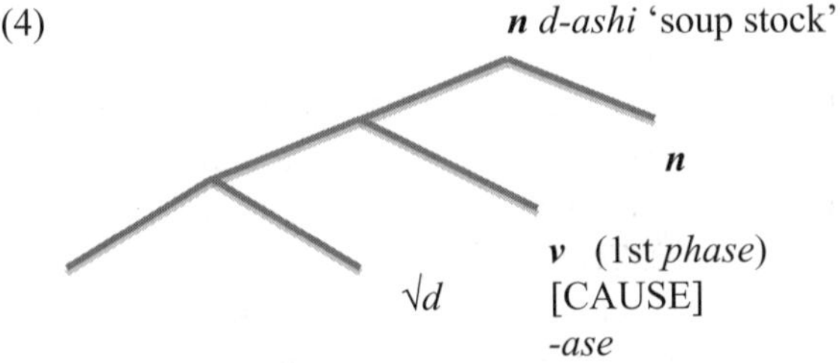

Given the strong support for identifying the phonologically overt affixes creating the causative and inchoative Japanese verbs as little *v*'s, Volpe finds that nominalizations built outside the little *v* head in Japanese using the "continuative" morpheme as in table 6.1 often involve special meanings of the root triggered across the little *v* head. That is, the meaning of such nominalizations is not compositional, given the meaning of the verbs. (Explaining the VI(s) for "continuative" and their ultimate phonological realization would take us beyond the limited scope of this chapter; the reader is invited to consult any standard reference grammar of Japanese for a description of the formation of the "continuative" and its various uses within Japanese.) Crucially, Volpe argues that not even the contrastive causative/inchoative meaning associated with the overt little *v* heads in these nominalizations is used or preserved in the meaning of the noun. For example, there is nothing about the meaning of *ko-e* "manure" that suggests the inchoative of "to become fat," nor is the meaning of *sag-ari* "hand-me-down" particularly compatible with the inchoative on which it is based morphophonologically, as opposed to the causative.

Table 6.1
Japanese Nominalizations from Volpe (2005)

Root	Verb-INTRANS	Verb-TRANS	Nominalization
√*ko(y)-*	*ko-e-(ru*-NON-PAST*)* 'to become fat'	*koy-as-(u*-NON-PAST*)* 'to fatten'	*ko-e* 'manure'
√*nag-*	*nag-e-(ru)* 'to flow'	*nag-as-(u)* 'to make flow'	*nag-ashi* 'a sink'
√*d-*	*d-e-(ru)* 'to exit'	*d-as-(u)* 'to expel'	*d-ashi* 'soup stock'
√*sag-*	*sag-ar-(u)* 'to be lowered'	*sag-e-(ru)* 'to lower'	*(o)sag-ari* 'hand-me-downs'
√*mag-*	*mag-ar-(u)* 'to bend'	*mag-e-(ru)* 'to bend'	*mag-e* 'a topknot, chignon'

If we take Volpe's description of the data as essentially correct, and the "continuative" ending in these nouns is actually conditioning a special meaning of the root (in the sense of an alloseme of the root), then two conditions would have to hold for these forms to obey the locality constraints on contextual allosemy supported in this chapter. First, in contrast to Volpe's analysis in (4), the continuative ending would need not to be, essentially, a little n head so that it would not constitute a phase head. To condition contextual allosemy on the root, it must fall within the same spell-out domain of the root, and thus not itself define its complement as a spell-out domain. Second, the little v head, although phonologically overt, must be semantically null or inactive. For the continuative to condition contextual allosemy on the root, it must be semantically adjacent to the root, without the v head getting in the way semantically.

Even without providing a technically complete analysis of these Japanese nouns, we can see how they in fact meet these two conditions. The continuative is an inflected form of the Japanese verb, used in a variety of constructions; it is "inflectional" rather than "derivational" in the sense that it does not by itself change lexical category. To conjugate a verb, you must know its continuative form—that is, the form is paradigmatic within the inflectional system of Japanese. The form of a verb in the continuative, then, might be compared to a participle in Indo-European languages.

And we have already seen evidence that the little v head does not contribute semantically to the meaning of the noun. These nouns are not essentially result nouns (they do not require, for their meaning, that an event took place) and, as argued by Volpe, they do not preserve in any way the meaning of the little v that distinguishes causatives from inchoatives, although their phonological spell-out does respect this distinction. (Some verbal roots build this type of noun from the causative, some from the inchoative, and a few build one noun from each alternant.) So the nouns behave as if the key semantics of little *v*—the introduction of an event variable, with a value of causative or inchoative—has been ignored in the meaning of the nouns. The crucial aspects of the Japanese examples that make them examples of local contextual allosemy rather than counterexamples to locality claims are (1) the absence of a verbal event meaning within the nominalizations and (2) a nonphasal ("inflectional") trigger for the construction.

Technically, we require a number of additional things to make this analysis work formally. First, it is necessary that the difference between causative and inchoative values of little *v* be syntactic, since this information is available for VI at the PF interface even when it is not used semantically (see the discussion in Marantz 2012). Second, the continuative head must be allowed to choose an alloseme of the little *v* head that is semantically null, as well as an alloseme of the root that creates in essence the meaning of a bound nominal root, wishing to combine semantically with a head that introduces an entity variable. So the whole continuative form of the verb must combine with a null little *n* head to yield a usable semantic value. As far as I can tell, these necessary ingredients for a contextual allosemy analysis of the Japanese nominalizations are all reasonable, but this broad promissory note for an analysis needs to be cashed in with a systematic analysis of a chunk of Japanese.

Again targeting the claims in Marantz 2001, Anagnostopoulou and Samioti (forthcoming) (hereafter, A&S) present another potential counterexample to the phase-based locality domain for contextual allosemy on a root, here from Greek. As with the Japanese example just discussed, A&S are building on and extending a long, deep, and convincing literature on Greek participles that lays the groundwork for their claims. Their paper reviews the literature on Greek stative participles, in particular on the distinction between stative participles in *-t-os* and those in *-men-os* (the *-o-s* pieces of the ending are case/number/ gender morphology of the adjectival/nominal sort that may vary to agree with a noun when the participles are used in verbal or adjectival contexts). In general, they argue, following previous analyses, *-menos* attaches to verbs preserving an event interpretation associated with the event variable introduced by little *v* while *-tos* attaches directly to roots, to create adjectives (or nouns) that do not imply an event.

There are interesting systematic exceptions to this general distinction between *-tos* and *-menos* that A&S review, but the center of their discussion relevant to the present chapter is a set of *-tos* participles that show the same syntactic and semantic behavior as *-tos* participles arguably built on roots (and not verbs) but that phonologically look like they are built on verbs with overt little *v* verbalizing morphology. In addition to not displaying the semantics (and associated syntax) of an embedded little *v*, these participles sometimes show special or unique meanings for their roots, meanings not associated with the verb outside the *-tos* construction. Examples of these stative adjectives, and their corresponding verb forms, are shown in (5). Here, the special meanings are not obvious, but the initially remarkable fact about the *-tos* participles is their co-occurrence with overt little *v* morphology, where stative *-tos* without eventive semantics usually attaches directly to a root morphophonologically.

(5) *Anagnostopoulou and Samioti, forthcoming.* -tos *statives (or nominals)*
    with special meanings
    a. axn-is-tos   'steaming hot' axn-iz-o 'steam'
    b. koudoun-is-tos 'ringing'   koudon-iz-o 'ring (a bell)'
    c. magir-ef-tos   'cooked'   magir-ev-o   'cook'

As A&S make clear, "Crucially, these verbalizers [in (5)] do not introduce an event variable and, therefore, *-tos* participles never have event implications (and the syntactic properties associated with an event interpretation)." Examples with more obvious root allosemy are shown in (6). Note that the "base" verb for the *-tos* participle need not exhibit a literal meaning as a participle, as in (6c).

(6) *Idiomatic interpretation of root-v-tos*
    a. kol-a-o   kol-i-tos   'close friend'
       glue-1sg   lit. glued
    b. xtip-a-o   xtipl-i-tos   'striking'
       bang, hit-1sg lit. whipped
    c. xon-ev-o   xon-ef-tos   'inside the wall'
       digest   no lit. meaning

A&S point to an interesting generalization about the roots in these apparent counterexamples to the claim that noneventive *-tos* participles are built from roots rather than verbs: the roots in these words are arguably nominal, adjectival, or uncategorized roots—they are not semantically verbal roots (event modifiers). The intuition is that the *-t-* participle morpheme requires a verbal stem syntactically.[9] At semantic interpretation, however, the semantics of the *v* node can be ignored and the *-t-* can create a meaning directly from the root.

We find in the Greek *-t-os* forms the necessary pieces to remove them from the list of counterexamples to the locality of allosemy on the root. The little *v* head that intervenes syntactically and phonologically between the trigger of allosemy, the participle *-t-* head, and the root is semantically null, not contributing an event variable to the construction, which then lacks the syntax and semantics of an event. In addition, the participle head itself is arguably not a little *a* or *n* head but rather part of the inflectional domain in Greek. So, the participle head would choose the null alloseme of the little *v* head, as well as determining the alloseme of the root that it attaches to directly, semantically. Combining with the root, it creates a derived root meaning, and thus a semantic bound root that will require a little *a* or *n* head to combine with. As in Japanese, we may assume that the actual category-determining head in the Greek constructions is phonologically null.

The structure of these almost counterexamples to the locality of contextual allosemy from Japanese and Greek is so similar that they point directly to where we might find a similar class of examples in English. What we would look for is an adjective or noun produced from a verb created with an overt little *v* (e.g., *-ize*) and overt participle morphology (*-ed* or *-ing*) where the meaning of the word does not include an event variable—that is, does not include the meaning usually associated with a little *v* head. Such cases are found in (7a–c). The examples in (7d–f) are provided for contrast; here the root is categorized by an overt little *a* head realized by *-al* phonologically. Since the input to the little *v* head is itself a typed constituent, the root is blocked from contextual allosemy triggered by the participle head over a potentially semantically null little *v* head.

(7) *English shows predicted "counterexamples" with stative passives outside* -ize *little* v *on "nominal" roots*
   a. quantized energy
   b. pulverized lime
   c. atomized individual
   d. globalized universe
   e. nationalized island
   f. fictionalized account

In (7a), *quantized energy* has a reading, "energy in quantum units," that does not imply any event of creating these units. Similarly, *pulverized lime* can mean "lime in powder form" without the implication of crushing. Finally, an *atomized individual* is one separated from society, but not necessary separated via any event of separation. In contrast, a *globalized universe* cannot mean a round universe or a universe that consists just of our world, but must be a universe that has been made global (not an easy reading to construct). Similarly, a *nationalized island* must be one that has been nationalized, not an island nation or an island of nations—perhaps an island taken by the government from private ownership. And a *fictionalized account* must have undergone a process of transformation from nonfiction to fiction; it cannot be simply a made-up story (not simply a fiction(al) account).

As in the Japanese and Greek cases discussed above, the head triggering the special meanings of the root here in the English cases is crucially not itself a category head. The stative passive *-ed* in (7) is quite parallel to the Greek *-t-*, and has also been argued to be able to attach directly to roots in stative passives with special meanings (e.g., in a "hung jury"). Arguably a participle head, the *-ed* in (7a–c) would choose a null semantic alloseme of the *v* head of its stem (avoiding any event interpretation) and attach semantically directly to the root, creating a derived bound root of an appropriate sort to modify a little *n* or little *a* head. So the stem created by the participle heads in (7a–c) would need a phonologically null *a* or *n* head to attach to to complete these forms syntactically and semantically. That is, *quantized* is not an adjective because the participle ending realizes an *a* head from the syntax but because the participle creates a bound stem that syntactically merges with an *a* head, which ends up being phonologically null.[10]

6.5 Conclusion

To review, I have argued here that the general phase-based architecture of the Minimalist Program with Distributed Morphology can be seen as imposing locality constraints on the context for contextual allomorphy and contextual allosemy. The facts as we see them are consistent with the "single-phase" hypothesis that the spell-out domains for the PF and LF interfaces are the same, defined by the same phase heads.

Each phase head sends its complement domain for interpretation. This includes previously interpreted phases, elements moved to the edges of those previous phases, and non–phase heads in the "extended projection" of a selected complement to the phase head. So, for example, the C phase head would send the as-yet-uninterpreted *v*P for interpretation, plus elements moved to the edge of the *v*P, plus modifiers of the *v*, like the verbal root, and the T, Neg, and aspect heads, if any, in the extended projection of *v*. Included within this domain would be already-spelled-out parts of the *v*P, in the complement domain of *v*(+voice), whose semantic and phonological values would be combined with the elements of the new spell-out domain.

In addition to the demand that the context for contextual allomorphy/allosemy be local to the spell-out domain, considerations of phonological and semantic composition lead to an additional constraint on local context: Only accessible features of a constituent that combines with a head *h* within the spell-out phase of *h* may serve as context for contextual allomorphy/allosemy, where "combines with" means phonological adjacency at PF and semantic adjacency at LF. So, phonological nulls will not get in the way of phonological adjacency (if they are part of the same spell-out domain as the constituents on either side) and semantic nulls will not get in the way of semantic adjacency. Since the two interfaces are connected only via the syntax, the phonological nullness of a constituent cannot be (directly) relevant to semantic interpretation, and we expect that phonologically non-null heads might be semantically null at LF and vice versa, leading to apparent exceptions to the claim that the locality domains for contextual allomorphy and contextual allosemy are the same.

I reviewed three cases from different languages where the proposed constraint on contextual allosemy appears to be violated. In each case, a morpheme that is arguably not itself a category head, and thus not a phase head, conditions allosemy on a root over a phonologically overt little v head. In each case, however, the semantics of this little v head are suppressed (nonexistent) within the derived word. This sort of situation exactly parallels the apparent long-distance contextual allomorphy in the case of the English past tense, where a non–phase head, the past tense, conditions contextual allomorphy of the root over a phonologically null little v head. The strong prediction of the theory of contextual allomorphy/allosemy proposed in this chapter is that all apparent counterexamples to the locality of the triggering context for the allomorphy/allosemy should conform to this pattern. In particular, the intervening phase head should be relevantly null (phonologically null for contextual allomorphy; semantically null for contextual allosemy) and the triggering head should not itself be a lexical category determiner—that is, not a phase head.

Notes

I would like to thank Ora Matushansky, Dave Embick, Heidi Harley, Tatjana Marvin, and Elena Anagnostopoulou for their comments on earlier versions of this work, as well as Allyson Ettinger, for editorial assistance. This work was supported in part by grant G1001 from the NYUAD Institute, New York University Abu Dhabi, and by grant No. BCS-0843969 from the National Science Foundation.

1. The name *Distributed Morphology* is from David Pesetsky, suggested at a meeting that Morris, David, and I had to discuss where morphology was going in the early 1990s. Originally, the term was intended to invoke current ideas of David, Hagit Borer, Mark Baker, and others that distributed the work of morphology and word formation across different places in the grammar. For example, Pesetsky (1985) argues that parts of words undergo QR at LF to establish the semantic scope of morphemes that start out as parts of words in the syntax, putting morphological matters outside the lexicon and into the syntax, broadly construed. Over the years, particularly in the wake of Halle and Marantz 1993, the term *Distributed Morphology* has narrowed in usage to refer to

the particular set of assumptions of Halle and Marantz 1993, 1994, and other work that takes Halle and Marantz 1993 as its starting point.

2. For present purposes, we could equivalently say that the allomorph of TEACH is /tɔt/ in the presence of past, with the past-tense allomorph being zero.

3. Tatjana Marvin points out that the inclusion of the root, little *v*, and some inflection (here the past tense) within the same spell-out domain seems in conflict with any easy equation of spell-out domain with the notion of "cyclic domain" required for phonology. In particular, there is evidence that the stem in English verbs is subject to cyclic phonological rules prior to the phonological inclusion of progressive *-ing* (cf. the classic contrast between the two-syllable pronunciation of the noun *lightening* and the three-syllable pronunciation of the progressive of the verb *lightening*, where the schwa in the verbal stem is arguably preserved via cyclic rule application, if the verbal stem constitutes a cycle). Bobaljik (2008) discusses evidence that the verbal root, but not the nominal root, constitutes a cyclic domain in a variety of language, accounting for phonological contrasts between verbs and nouns. Within the present framework, this pattern of data could indicate that spell-out domains and phonological cycles should simply not be equated; sometimes a single spell-out domain might contain multiple cycles, where a phonological cycle is determined by phonological properties of VIs and of structures of morphemes. However, further research may find that there is additional structure in the relevant cases for verbs that induces a spell-out domain just where a phonological cycle is motivated.

4. The contrast between phase heads and non–phase heads in their ability to trigger contextual allosemy on the heads of their complements will be crucial in the examples to be discussed below. In particular, in the case of the Japanese, Greek, and English examples, a non–phase head will cause a little *v* head within its complement domain to receive a null interpretation, with the result that there is no event interpretation present within the derived words. The theoretical apparatus of this chapter predicts that a phase head, say a category node, could not choose a null alloseme of a little *v* head, restricting the phenomena discussed below to situations in which a non–phase head is involved in creating nouns and adjectives.

5. The plural of *house, houses,* also contains the contextual allomorph of the stem with the voiced fricative. Within the current framework, this fact suggests that plural is a non–phase head similar to past, with irregular plurals and irregular past tenses receiving similar treatment. In this context, we should note that the present theory does *not* predict that the plural of *computer mouse* should necessarily be *computer mouses* as opposed to *computer mice*. If (computer) *mouse* is formed from the noun *mouse*, with an additional little *n* head, then the plural should be *mouses*, parallel to the past tense of a verb *house* formed from the noun *house* being *hou[st]*. However, the computer-input-device meaning of *mouse* could involve contextual allosemy of the root *mouse* in the environment of little *n*. In that case, the irregular plural *mice* would still be available, given the presence of the root, the *n* head, and plural in the same spell-out domain and the phonological emptiness of the VI realizing the *n* head. Variation between *mice* and *mouses* in discussions of computers could then involve structural ambiguity in the analysis of (computer) *mouse*.

6. See Simon, Lewis, and Marantz 2012 and the references cited there. I should note here that different researchers and different research traditions make different cuts

among "polysemy," "homophony," and "vagueness." Here I try to clarify what's crucial for the present discussion in distinguishing homophony and polysemy by providing specific relevant examples.

7. Apparently some speakers also allow the "sphere" reading of *globe* in *global*, making *global* a synonym of *globular*. For these speakers, the little *a* is not choosing an alloseme of the stem, and so both meanings are carried up to the adjective. However, the "sphere" reading is not preserved by *-ize* in *globalize* for these speakers either, and so cannot be revived in *globalization*.

8. Heidi Harley brings up the interesting case of *hospitality*. Here, from a root *hospit-*, English creates both a noun, *hospital*, and an adjective, *hospitable*. *Hospitality*, which seems to involve the *n*-forming *-ity* attaching to *hospital*, does not preserve the contextual meaning of the root found in *hospital*, but does seem to be related to the meaning of the root in *hospitable*. We know that *-ity* likes to attach to adjectives in *-al* and *-able* but otherwise generally attaches to roots (see Embick and Marantz 2008). If *hospital* were an adjective formed with *-al*, *hospitality* would be a counterexample to the claims of this chapter, since *-ity* would seem to be choosing an alloseme of *hospit-* (the one also chosen by *-able* in *hospitable*) over the adjective-forming *-al*. However, *hospital* is a noun and thus arguably not the adjective base of *hospitality*. We seem here to have a bound root *hospital-* that is perhaps an allomorph of the root *hospit-* in *hospitable* and *hospital*, with the same alloseme as it has in *hospitable* but not *hospital*.

9. Harley suggests the possibility that the semantically vacuous, but phonologically overt, *v* morphemes in Japanese and Greek might be inserted phonologically, rather than syntactically, along lines suggested for *do*-insertion in English. Given the current framework, the fact that the spell-out of the little *v* morphemes is contextually determined (roots pick different verbalizing suffixes) indicates that what is inserted must be a morpheme in the syntax, rather than a VI in the morphophonology, and morphemes are organized by syntactic structure, prior to phonological spell-out. Embick and Noyer (2001) in fact argue that *do*-insertion itself must be syntactic.

10. Here it is worth reminding the reader that the ability of a non–phase head to combine with a root to yield a special meaning over a syntactically active but semantically null category head does not predict that past-tense forms in English might have special meanings not apparent for present-tense forms of the same verbs, parallel to irregular past-tense forms showing phonologically special forms over a (phonologically null) *v* head. If a past-tense verb is to have a past-tense meaning, it must combine semantically with an event variable; therefore, a past-tense verb must have a semantically active *v* node and tense (with a tense reading) may not combine directly with the root.

7 Cycles, Vocabulary Items, and Stem Forms in Hiaki

Heidi Harley and Mercedes Tubino Blanco

7.1 Introduction

The analysis of arbitrary morphological classes has a number of architectural implications in Distributed Morphology (Halle and Marantz 1993). There is no central repository of Saussurean 'words' in the framework—no sound-meaning pairings that are the building blocks for both phonological and semantic sentence-level representations. Instead, there are separate lists. One list contains all syntactic and semantic information necessary for the derivation of a well-formed LF representation, and forms the input to the syntactic derivation. A second list, the Vocabulary, describes the phonological realizations that are inserted as exponents of particular syntactic terminal nodes, following all syntactic operations. This raises the question of where class features are located. What elements do rules which are sensitive to class membership refer to? Are they sensitive to properties of the abstract syntacticosemantic formatives of the first list? Or are they instead sensitive to properties of the phonological exponents, the Vocabulary Items in the second list?

In Distributed Morphology, any features which are syntactically (and possibly semantically) active must be a property of the abstract morphemes which are input to syntactic derivation. Embick and Halle (2005) treat even Latin conjugation class features in this fashion, attaching them to the roots in the first list, input to the syntax.

In contrast, we will argue that class features in Hiaki are not properties of roots in the syntax but rather are properties of Vocabulary Items, the phonological exponents inserted at the end of the syntactic derivation. Irregular morphophonological rules (readjustment rules) apply to a particular class of Vocabulary Items in the appropriate morphosyntactic environment. Classifications of this kind play no role in the syntactic/semantic computation, but are crucial in triggering the application of the appropriate morphophonological rule to yield the correct surface form in such cases. The existence of such morphophonological classifications, irrelevant to syntax, is thus an argument against the lexeme, as such, and in favor of the DM-style separation of the two lists: List 1, input to the syntax (the source of the Numeration in Minimalist syntactic theories), and List 2, Vocabulary Items which simply realize the output of the syntax. Further, the Hiaki case provides a clear argument for Vocabulary Insertion applying to Root elements (l-morphemes, in Harley and Noyer 2000's terminology), as well as to f-morphemes. In addition, the notion of a phasal cycle within the word proves useful in permitting a simple statement of the relevant conditioning context for the application of morphophonological rules.

The Hiaki case presents many of the same morals for the architecture and the notion of "stem" as the Latin perfect does, as elucidated in Embick and Halle 2005, but some of the issues arise in even starker relief due to the cross-categorial nature of the stem classes, and especially due to the interaction of the stem classes with suppletion, in particular with the suppletive roots of Hiaki.

7.2 Hiaki Stem Classes

In Hiaki (Yaqui), lexical stems have bound and free alternants. The bound forms are used as the base for affixation of (broadly speaking) derivational morphology, while the free forms are the base for affixation of (broadly speaking) inflectional morphology, and also of course can stand alone, without any affix. The bound and free stems for *poona* 'play, beat' and *kiima* 'bring (PL. OBJ)' are illustrated in (1), the latter affixed with a derivational suffix.

(1)   *Free*   *Bound*

    a. **poona**   **pon**-tua

       play   play-CAUS

       'is playing'   'is making (someone) play'

    b. **kiima**   **kima'a**-*tua*

       bring.PL   bring.PL-CAUS

       'is bringing (things)' 'is making (someone) bring (things)'

An almost-complete list of the free-stem-selecting and bound-stem-selecting suffixes are listed in (2a) and (2b) respectively:

(2)

Bound-stem and free-stem suffixes of Hiaki (Harley and Tubino Blanco 2010)

a. *Hiaki verbal suffixes that require the bound stem (in no particular order)*

   -tua (CAUS)   -'ea (DESID) -su (COMPL) -se/-vo (go) -pea (DESID)

-tevo (ICAUS) -ri (OBJ.PPL)   -la (PPL)   -taite (INCH)   -'ii'aa (DESID)

-ria (APPL)   -tu (become)   -ri (PPL)   -naate (INCH)   -hapte (INCH)

-roka (QUOT)   -vae (PROSP)   -le (consider) -wa (PASS)   -ne (IRR)

-yaate (CESS) -siime (go along) -sae (DIR)   -na (PASS.IRR)

b. *Hiaki verbal suffixes that require the free stem (again in no particular order)*

-k (PERF)   -ka (PPL)   -n (P.IMPF) -kan (PST.IMPF) -o (if/when)

-'u (OBJ.REL) *me* (SUBJ.REL)

Note that a single word can and often does contain both derivational and inflectional suffixes. In such cases, the derivational suffixes occur closer to the stem than the inflectional one (as expected) and the bound-stem form is chosen, as in, for example, the perfective form of the causative of (1a), *pontua-k* 'play-CAUS-PERF' (i.e., 'made (someone) play').[1]

While several of the free-stem suffixes are simple coda consonants, others are CV or CVC syllables; many of the bound-stem suffixes are also CV syllables, so it seems unlikely that the choice of bound or free stem is driven by prosodic requirements. In addition to the suffixes listed in (2b), the bound-stem form is also required for the nonhead member of any compound (verbal or nominal); see section 7.3.2 for exemplification and discussion.

Harley and Tubino Blanco (2010) describe the various patterns of bound-stem formation in verbs, distinguishing three primary stem classes and several subclasses and irregularities. Classes are distinguished according to the shape of the bound stem and the nature of its relationship to the free stem. We summarize these below.

The three primary classes, comprising dozens of exemplars each, are distinguished by whether the stem form is a truncated version of the free form (Class 1), an augmented version containing an extra syllable of the form -$\mathit{?V}$, where the vowel is a copy of the final vowel of the stem (Class 2), or identical to the free form (Class 3). Each type is exemplified in table 7.1.

It is important to note that vowel shortening in the stem forms is a regular process in the language, which usually occurs quite generally, under any kind of affixation, to free as well as bound stems. A change from long to short vowel between free and bound forms, then, is not part of the stem-formation process, but rather a regular phonological rule of the language triggered by any kind of affixation.

Table 7.1
Major stem-alternation classes of Hiaki

Class 1: Truncation			Class 2: Echo-vowel		
Free	Bound	English	Free	Bound	English
a. *poona*	*pon-*	'pound'	a. *bwasa*	*bwasa'a-*	'cook (TR)'
b. *miika*	*mik-*	'give'	b. *kiima*	*kima'a-*	'bring (PL)'
c. *bwase*	*bwas-*	'cook (INTR)'	c. *yoore*	*yore'e-*	'heal'

Class 3: Invariant		
Free	Bound	English
a. *kivacha*	*kivacha-*	'bring (SG)'
b. *hamta*	*hamta-*	'break'
c. *suua*	*sua-*	'care.for'

There are several minor subclasses of stems. Three of these minor classes are listed in table 7.2. Another group of subclasses contains free forms which end in two distinct vowels (usually separated by a glottal stop, but sometimes not, or optionally so). Their corresponding bound stem is created by changing the second of the two vowels to match the first; a glottal may appear in the bound form even if it is absent in the free form (table 7.3). In addition to these groups of forms, there are a number of effectively irregular forms, whose bound stems are not related to their free forms according to any of the patterns exemplified in the accompanying tables, or to any other pair that we know of. A sampling of these unique cases is given in (3).

(3) *Irregular free~bound pairs*
    a. *yepsa* → *yevih-* 'arrive (SG)'
    b. *suulu* → *suluu-* 'slide, slip'
    c. *sevea* → *seve-* 'catch a cold, get cold'
    d. *suawa* → *suan-* 'be watched'
    e. *kepe* → *kup-* 'close one's eyes'
    f. *hia* → *hiu-* 'vocalize'
    g. *ho'otia* → *ho'otiu-* 'snore'
    h. *ve'a* → *vi'i-* 'save, reserve'

An important fact to note about all the bound-stem forms, of whatever class, is that they are phonologically related to their free forms. Consonantal material is not affected by the alternation, with one or two types of exceptions (the general *s*→*h* / ______ C pattern, and the restricted *v*→ *p* and *ch*→*t* alterna-

Table 7.2
Minor stem alternation subclasses

Subclass i: *-e → -i*			Subclass ii: *e'e → -i'i-*		
Free	Bound	English	Free	Bound	English
a. *hamte*	*hamti-*	'break (INTR)'	a. *he'e*	*hi'i-*	'drink'
b. *chihakte*	*chihakti-*	'smash'	b. *ne'e*	*ni'i-*	'fly'
c. *yu'e*	*yu'i-*	'undo'	c. *ye'e*	*yi'i-*	'dance'
d. *vuite*	*vuiti-*	'run (SG)'	d. *che'e*	*chi'i*	'suckle'

Subclass iii: *-u → -oe*		
Free	Bound	English
a. *kiimu*	*kimoe-*	'bring (SG)'
b. *vaasu*	*vasoe-*	'soak'

Table 7.3
Vowel-copying subclasses

Subclass iv: *-o'a/-oa → -o'o-/-oo-*			Subclass v: *-e'a → -e'e-*		
Free	Bound	English	Free	Bound	English
a. *hi'ivoa*	*hi'ivoo-*	'cook'	a. *eo'ote'a*	*eo'ote'e-*	'be nauseated'
b. *hovoa*	*hovo'o-*	'get full'	b. *ea*	*ee-*	'feel'
c. *ko'a*	*ko'o-*	'chew'	c. *me'a*	*me'e-*	'kill (SG)'
d. *to'a*	*to'o-*	'pour, lay down (PL)'			

Subclass vi: *-a'e → -a'a-*			Subclass vii: *-u'e/-u'a → -u'u-*		
Free	Bound	English	Free	Bound	English
a. *bwa'e*	*bwa'a-*	'eat'	a. *nu'e*	*nu'u-*	'get, acquire'
			b. *yu'a*	*yu'u-*	'push'

tions). Similarly, the vowel in the first syllable of the stem is not affected by the free/bound alternation, the only exceptions being the *e→i* alternations in subclass (ii) and (3h), and the irregular *kepe→kup-* pair in (3e). To the best of our knowledge, there are no suppletive bound-stem forms in the language—there is no case in which the bound form is phonologically unrelated to the free form. (There is considerable suppletion in the language, as discussed in section 7.4, but it is never conditioned by the bound-stem/free-stem alternation.)

Below, following Embick and Halle (2005), we will build an argument from these Hiaki facts against the listing of stem forms in Hiaki. Rather, we will argue that the bound-stem forms should be derived via the application of readjustment rules—phonological rules restricted to apply to only a given group of Vocabulary Items, which apply following Vocabulary Insertion— rather than by competition for exponence of a given Root node.

7.3 Listedness vs. Readjustment: Stems in Distributed Morphology

We now turn to a discussion of the theoretical implications of the Hiaki free- and bound-stem forms. First we review the conclusions of Embick and Halle (2005) concerning the superflous nature of the notion of 'stem' in Distributed Morphology. We then present two arguments against the notion of listing of stem forms in Hiaki, along the lines of the argumentation Embick and Halle (2005) present against the proposal of Aronoff (1994) concerning the listing of Latin verb-stem forms.

7.3.1 On the Nonlistedness of Stems in DM: Embick and Halle (2005)

Embick and Halle (2005) provide an extended discussion of the status of the notion "stem" in Distributed Morphology. In particular, they reprise the argument against the position of Anderson (1992) according to which different but phonologically related stem forms such as *sing~sang* are listed, competing with each other for exponence in the same way that distinct, listed suppletive/ allomorphic forms such as *go~went* or *-ed~-t* do.[2] Their conclusion, echoing the discussion of Marantz (1997) and Halle and Marantz (1993), is that such alternations are substantially different from suppletive cases, the latter being truly rare within and across languages. If the *sing~sang* alternation is treated via listing and competition, exactly as suppletion is treated, the theory itself imposes no principled distinction between suppletion and restricted but basically phonological alternations; the result is simply a cline from maximal irregularity to complete irregularity. In contrast, if the vowel change in *sing~sang* is implemented by a morphophonological readjustment rule, it is expected that such alternations should behave in accordance with normal phonological patterns, which, by and large, they do.

Halle and Embick propose a treatment of Latin verb formation in response to a stem-storage model proposed by Aronoff (1994). In their account, a particular set of readjustment rules are triggered in the environment of a certain exponent of the Asp node, thus explaining the appearance of a selectional relationship between a particular stem form and those aspect exponents. A truly selectional effect would require the listing of the stem form (so it could be selected for), but given the independent necessity of readjustment rules, the fact that the same effect can be captured via a restriction on a readjustment rule means that the positing of a listed stem form to account for these cases is unmotivated.

We can easily imagine how a "listing" account of Hiaki verb-stem forms would work. The free form of a given Hiaki verb and its bound form would each be listed as alternate phonological forms for a single lexeme (or element from List 1, in DM terms). One would be the Elsewhere form, inserted when no more specific criterion is met. Although it might seem natural to assume that the free form is the Elsewhere form, in fact it is easier to specify a unified environment for occurrences of free Hiaki verb stems than for bound ones: basically, the free form occurs whenever the verb is inflected, which we could characterize as "whenever V is immediately adjacent to Asp^0" (which might, of course, have a null realization, as in the present tense, producing the unaffixed free forms). The bound stem, then, would be the elsewhere form, occurring any time the verb was not adjacent to Asp^0, which in fact will be exactly the set of environments in which it is affixed with derivational morphology. To illustrate how such a "listing" analysis would play out, Vocabulary Insertion rules of this kind for *poona* 'play.instrument' (Truncation class), *bwasa* 'cook' (Echo Vowel class), and *chihakte* 'smash' (e→i class), as well as the unique bound stem of *hia* 'sound, vocalize', are given below. An Invariant class item like *sova* 'roast' would only have one listed form:

(4) a. [PON] $_V$ 'play.instrument, strike' → *poona* / ______ Asp^0
         pon   Elsewhere

 b. [BWASA'A]$_V$ 'cook' → *bwasa* / ______ Asp^0
         bwasa'a Elsewhere

 c. [CHIHAKTI]$_V$ 'smash' → *chihakte* / ______ Asp^0
         chihakti Elsewhere

 d. [HIU] $_V$ 'sound' → *hia* / ______ Asp^0
         hiu   Elsewhere

 e. [SOVA]$_V$ 'roast' → *sova*

We will see that such an account suffers two significant drawbacks, described in the next section.

7.3.2 The Nonlistedness of Hiaki Stem Forms: Two Problems

The "listing" account of Hiaki stem alternations described above suffers from two substantial problems, the first of which is no doubt immediately obvious to the reader. The first problem is lack of insight into the morphophonology of Hiaki, the arbitrariness issue discussed by Embick and Halle (2005). Listed forms need not bear any relationship to their other alternant or to each other. There is no reason why they should fall into the general classes described in section 7.2, which are characterizable in broadly phonological terms (Truncation, Echo Vowel, etc.). Suppletive stem alternants should be possible, but do not occur. Positing stem classes is motivated by such groups of forms, but the listing approach does not reflect such groupings. In fact, the notion of "class" drops away entirely in such an account; the fact that a substantial group of verbs have an echo vowel at the end of their bound stem would be a simple coincidence. Indeed, there is considerable evidence supporting the psychological reality of synchronic rule application in irregular morphophonological classes in other languages (see, e.g., Yang 2002; Stockall and Marantz 2006). We assume that these Hiaki classes should be analyzed similarly to rule-based morphological classes in better-studied languages.[3]

The second problem arises from the fact, alluded to but not exemplified above, that the bound versus free alternation is a property of nouns as well as verbs in Hiaki, and that similar morphophonological processes derive bound nominal stems; bound nominal stems seem to fall into the same general classes that bound verbal stems do. Hiaki does not have as much robust derivational morphology for nouns as it does for verbs, but nouns can undergo incorporation or compounding productively, and when they do, they occur in their bound form. Consider the examples below.

(5)   *Free*   *Bound*

  a. ***mama-m***[4]   ***mam***-*pusiam*   Truncation
      hand-PL   hand-eyes
      'hand'   'fingers'

  b. ***chiiva***   ***chiva'a***-*tu*   Echo Vowel[5]
      'goat'   goat-TU
                           'billy goat'

  c. ***avaso***   ***avas*** *naawa*   Truncation
      'cottonwood'   'cottonwood root'

  d. ***hi'u***   ***hi'u***-*se*   Invariant
      'greens'   green-V
                           'collect greens'

These patterns are robustly attested across the nominal vocabulary as well as in verbs. That is, all roots, not just verbal roots, have bound- and free-stem forms. The problem this creates for a "listing" approach is that the contexts that condition the appearance of each form are difficult to state in a unified way, although descriptively, the contexts are clearly unified: inflectional morphology, whether verbal or nominal, attaches to free forms; derivational morphology attaches to bound forms. These are not the typical kinds of conditioning environments for insertion rules, however. The rule above referred to Asp^0 as the conditioning context for the appearance of the free form, but this is obviously inadequate to characterize the nominal stem alternations in (5), which do not appear in a verbal extended projection. To unify the conditions for the insertion of the free form across nominal and verbal environments, a diacritic, perhaps something like [+inflectional], would be required, since it would be different inflectional categories, Asp^0 in the case of verbs and probably Num^0 in the case of nouns (cf. (5a)), which trigger the insertion of the free form.

Further, and more theory-internally, there is a problem of locality in stating the conditioning environment. In DM, the fundamental lexical terminal nodes are acategorial, provided with their nominal or verbal character by merger with an n^0 or v^0 head, which, in a "listing" approach, intervenes between the $\sqrt{}$ element being inserted and the supposedly conditioning inflectional category. (The acategorial character of roots is supported by the fact that nouns and verbs are parallel in having distinct stem forms with similar formation patterns.) The theory does not make available any obvious notation that will ensure that just a *single* x^0 category intervenes between a terminal node and the inflectional category which licenses the insertion of the free stem. A rule like that in (4) demanding adjacency between a $\sqrt{}$ and some head with a [+inflect] diacritic is unworkable even for underived nouns or verbs, since the v^0 or n^0 intervenes between the $\sqrt{}$ and the [+inflect] head above, so the free forms would never meet the licensing criterion. Both simple nouns and verbs and those that undergo further derivation require an x^0 to be adjacent to the $\sqrt{}$, and so the contexts for $\sqrt{}$ spell-out are not differentiated locally.[6] A stem-insertion rule would have to make reference to a complex hierarchical context like [[_______ x^0] $X^0_{[+infl]}$] in order to even get the facts descriptively correct.[7]

To recap: The first problem for a listing approach is the fact that there are groups of lexical items whose free and bound forms appear to be related by particular phonological processes. Positing stem classes is motivated by such groups of forms, but the listing approach does not reflect such groupings. The second problem is the difficulty of stating the correct morphosyntactic conditioning environment for insertion of particular listed items, which crosscuts lexical categories: insertion of the free form would have to be triggered by diacritics such as [+inflect] occurring in a specific but nonlocal configuration with respect to the conditioned root.

Given that the notion of "stem-formation class" is appropriate in the characterization of the Hiaki data, and that a listing approach fails to capture this notion, what would a more successful proposal look like? We turn to this question in the next section.

7.4   The Conditioning of Stem-Forming Readjustment Rules in Hiaki

Having dismissed the "listed-stem" approach, we next consider an alternative more consonant with the conclusions of Embick and Halle (2005), according to which the free/bound stem alternations are derived via morphophonological readjustment rules conditioned by classes, aka lists of roots (which, in the case of wholly irregular stem-forming rules, may just contain a single item). We briefly sketch how the problem of stating the correct environment for the application of such readjustment rules evaporates when we consider where the morphosyntactic cycle, aka phase boundary, is located within the Hiaki word. We then turn to our primary concern, which arises from the interaction of suppletion with Hiaki stem alternations. We argue that the fact that distinct suppletive realizations of the same root can belong to different form classes shows that the stem-class features of Hiaki words are a property of Vocabulary Items, not a property of the underlying abstract roots.

7.4.1   Readjustment Rules for Hiaki Stem Forms

Readjustment Rules apply to intermediate phonological representations during the derivation from syntactic structure to phonological form, following Vocabulary Insertion but prior to and consequently possibly bleeding the application of the regular phonological processes. The sequence of operations that may occur following *spell-out*, in the morphological component that maps syntactic representations to phonological objects for interpretation by the sensorimotor system, is illustrated in (6):

(6) *Operations that may apply to a syntactic representation at spell-out*

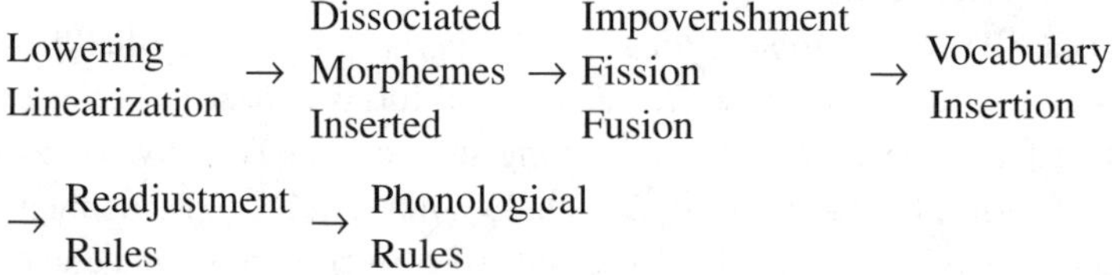

If Hiaki bound-stem forms are the result of applying Readjustment Rules to free forms, they would involve phonological processes like prosodically driven truncation (Class 1), affixation of CV syllable whose segmental content is determined by (e.g.) left-to-right spreading (the echo vowel, Class 2), raising of a final midfront vowel (Subclass i),[8] and so on. It is slightly more problematic to derive the free forms from the bound forms, mainly because the final material of the free forms appears to be more idiosyncratic than the final material of the bound forms, which often seems to be generated by copying the preceding vowel. So, for example, in Subclasses iv–vii, listed in table 7.2, the bound form always involves two copies of the penultimate vowel of the free form, but the ultimate vowel of the free form can be either -*a* or -*e*, apparently an idiosyncratic property of the stem.[9] Similarly, in Class 1 (the Truncation class) the bound forms lack material that is idiosyncratic to the free stem, not independently predictable in any way, so, for instance, there are truncated stems whose free form contains a final -*a* and the truncation consists in omitting that -*a* (*poona~pon-* 'play (instrument)', *bwiika~bwik-* 'sing', *miika~mik-* 'give') but other members of the class involve omitting a final -*e* (*bwase~bwas-* 'cook', *chupe~chup-* 'finish', *koche~kot-* 'sleep') or -*o* (*avaso~avas-* 'cottonwood'). A few involve omission of a final CV syllable (*hapte~hap-* 'stop') or two final vowels (*yosia~yoh-* 'sleep').

Given that the bound forms seem to be more regular and in some cases to have lost information present in the free form, we will take the free form to be basic and assume that the readjustment rules derive the bound form from the free form. We will not attempt to formalize each of the rules here, but merely insert descriptive names for them which summarize their general effect.[10] A few generalizations can be made; however, obviously much work remains to be done to flesh out the analytic sketch presented here.

First, class membership cannot be fully predicted based on the phonological properties of the free form. Near minimal pairs exist—for example *nooka* 'talk, speak to' is a member of the truncating Class 1 (bound stem *nok-*), but *yooka* 'paint' is a member of the Echo Vowel Class 2 (bound stem *yoka'a-*). Similarly, *naate* 'begin, start' is a member of the invariant Class 3, but *waate* 'remember' is a member of the *e→i* Subclass i.

Second, the bound form of a Truncation class word is always a heavy syllable, usually C(C)VC (as in *nok-* 'talk'). This leads to a conjecture regarding the derivation of the forms of subclasses iv–vii in table 7.3. It could be that two separate processes apply to generate this apparent pattern of vowel matching: Truncation (as in Class 1) followed by Echo Vowel (as in Class 2), with the latter possibly motivated by a preference for a bimoraic foot at the end of a bound stem.[11] This sequence would generate the bound form of, for example, *bwa'e* as follows:

(7) *bwa'e* + Truncation → *bwa*   →   *bwa* + Echo Vowel   → ***bwa'a-***

Such a derivation, if motivated, would provide additional support for the readjustment rule approach, since an individual lexical item could be on the lists for two (or more) of the readjustment rules (which would have to be ordered with respect to each other).[12]

7.4.2 The Domain of Stem-Forming Readjustment Rules

How can we characterize the environment in which such Readjustment Rules would apply? A key observation involves a close examination of the class of suffixes which require bound stems to attach to, given in (2a) above. Considered in terms of the hierarchical functional projections, the bound-stem requirement appears to end with affixes which occur in VoiceP. An overall templatic sketch of the left-to-right organization of the Hiaki verb complex is given below:

(8) *Schema of Hiaki verb structure: Everything except the second instance of V is optional*

   (dir/adv particle) (obj.clitic) = (incorp.N) - (RED) - (V) - V -
      (bound-stem suffixes) - **(PASS(+IRR))** - (free-stem suffixes)

Passive and future/irrealis markers are a rigid boundary between bound-stem affixes and free-stem affixes—between derivation and inflection. They themselves take bound stems. The irrealis suffix *-ne*, although not obviously Voice-related, actually encodes both irrealis and active Voice. It is in complementary distribution with the realis passive *-wa*; one cannot add irrealis *-ne* to a passive *-wa* (or vice versa) to mark a passive as irrealis. Irrealis passives employ a special portmanteau suffix *-na* in this position, expressing both properties at once. We speculate that Hiaki requires fusion of Voice and an immediately c-commanding Mood head. The key thing is that each clause can contain only one element that encodes Voice status, and that element must occur at the boundary between bound-stem-selecting and free-stem-selecting material.

This distribution is extremely suggestive from the perspective of modern syntactic theory. Chomsky (1995, 2001) suggested that the derivation be passed on *to Spell-out* in regular cycles, at specific boundary nodes, or *phases*. The external-argument-introducing node, which he termed v*, but which could equally be termed Voice, is one such phase. The fact that bound forms are required inside the passive implies that the stem-forming Readjustment rules apply *phase-internally*. Any element not left-adjacent to a phase edge is subject to the stem-forming Readjustment Rules. See Harley and Tubino Blanco 2012 for a more in-depth discussion of the syntactic character of this domain.[13]

It is worth noting that this view of the conditioning context entails that there are phase edges inside the nominal projection as well, which has become a relatively common position since the first introduction of phases (see the discussion in, e.g., De Belder 2011; Kharytonava 2011). Nominal compounding and derivation take place within that phase boundary and require bound nominal stems to appear everywhere except in final position; affixes in the higher inflectional projections of N (NumP, CaseP) are outside this phase boundary and hence attach to free nominal stems.

7.4.3 Class Membership Is a Property of Vocabulary Items, not Roots

Embick and Halle (2005) assume that the List 1 Root formatives are specified with phonetic feature information—that is, that Roots are not subject to late insertion from the Vocabulary (List 2), unlike functional morphemes. Their characterization of the content of Root elements is quoted here:

Roots: These make up the open-class vocabulary. They include items such as √CAT, √OX, or √SIT, which are sequences of complexes of phonetic features, along with abstract indices (to distinguish homophones) and other diacritics (e.g. class features). (p. 38)

Harley (2011) argues that Root nodes cannot contain phonetic features, but instead are subject to late insertion from List 2, the Vocabulary, just as abstract f-morphemes are. The argument is based on the claim that Roots have suppletive realizations that can compete for exponence, a claim supported by evidence from Hiaki and other Uto-Aztecan languages, as well as from other languages around the world, which have small but robust groups of suppletive forms that seem unlikely to be characterizable as light verbs. (This conclusion is also supported by evidence from other languages described in Bonet and Harbour 2012.) Harley concludes that List 1 Root elements are individuated solely by their indices.

Setting aside that particular issue here, we will focus on whether class features are diacritics on elements of List 1 or List 2. The interaction of suppletion and class features in Hiaki indicates that stem-class membership is a property of List 2 elements, not a diacritic on List 1 elements, as proposed by Embick and Halle. As we will see, this conclusion follows from the fact that suppletive variants of a single List 1 element can belong to different stem classes.

Hiaki suppletive verbs are conditioned by subject number (if intransitive) and object number (if transitive). The subjects of such verbs are argued by Harley, Tubino Blanco, and Haugen (2009) to be internal arguments; if so, conditioning of the relevant form is dependent on the number of the verb's internal argument. A representative sample is presented in (9):

(9) *Hiaki suppletive verbs*
    a. *vuite*   run.SG   *tenne* run.PL
    b. *siime*   go.SG   *saka*   go.PL
    c. *weama* wander.SG   *rehte* wander.PL
    d. *kivacha* bring.SG.OBJ *kiima* bring.PL.OBJ
    e. *vo'ote* lie.SG   *to'ote* lie.PL
    f. *weye*   walk.SG   *kaate* walk.PL
    g. *mea*   kill.SG.OBJ   *sua*   kill.PL.OBJ

The above represents a selection from a set of about fourteen or fifteen total suppletive verbs in the language; the particular set varies somewhat between speakers and across dialects, but the seven listed are among those that are consistent. It should be clear that it is implausible to treat these alternations in any other way than by listing. As in the case of the *go/went* alternation in English, the two forms compete with each other for realization, the winner determined by the number of the internal argument. A sample of such a rule is illustrated in (10). On the left is the abstract node from List 1, identified by its English translation;[14] on the right are the two Vocabulary Items competing to realize it.

(10) *Vocabulary Insertion rule for a Hiaki suppletive verb*
    RUN $\rightarrow$ /vuite/ / [[DP$_{+sg}$] ______ $\sqrt{}$]^{15}
          $\rightarrow$ /tenne/ elsewhere

If stem class were a property of the abstract node of List 1, then one would expect that the stem forms of both suppletive variants should belong to the same class and be subject to the same readjustment rules. Although this is the case for some of these suppletive pairs, it is not the case for others. Consider table 7.4.

From the data in table 7.4 it should be clear that suppletive variants of the same verb can have different stem-class properties. Therefore, stem classes are a property of the Vocabulary item, not of the abstract morpheme.

Note that the bound forms in table 7.4 support the point made earlier, that the bound forms are always related phonologically to the free forms. The bound form of a suppletive variant of a verb is never itself another suppletive variant; rather, it is a form that relates to the free form according to some reasonable phonological transformation.

The stem-form classes of Hiaki, then, are defined by the lists of roots that the context-restricted phonological readjustment rules apply to. The lists themselves are references to the phonological content of a Vocabulary Item from List 2, not to the abstract content of a List 1 item.[16]

Table 7.4
Stem classes of Hiaki suppletive verb alternants

	#	Free form	Bound form	Stem class
RUN	SG	*vuite*	*vuiti-*	$e{\to}i$
	PL	*tenne*	*tenni-*	$e{\to}i$
GO (PRES)	SG	*siime*	*sim-*	Truncate
	PL	*saka*	*saka'a*	Echo Vowel
WANDER	SG	*weama*	*wee-*	Truncate/EV
	PL	*rehte*	*rehti-*	$e{\to}i$
BRING	SG	*kivacha*	*kivacha-*	Invariant
	PL	*kiima*	*kima'a-*	Echo Vowel
LIE	SG	*vo'ote*	*vo'ote-*	Invariant
	PL	*to'ote*	*to'ote-*	Invariant
WALK	SG	*weye*	*wee-*	Truncate/EV
	PL	*kaate*	*kat-*	Truncate
KILL	SG	*mea*	*me'e-*	Truncate/EV
	PL	*sua*	*sua-*	Invariant

7.5 Conclusions

The Hiaki stem-form classes, then, support three important conclusions about
the architecture of the mapping to phonological structure. First, readjustment
rules (or their equivalent) are a necessary feature of any morphological theory
that wishes to characterize the patterns of stem formation of Hiaki: they are
clearly phonological (even prosodic) in character, and yet are constrained to
apply only to idiosyncratically restricted lists of lexical items. Second, the
domain of application of these rules lines up quite well with the Minimalist
construct of *phase*, supporting the notion that the derivation proceeds cyclically, phase by phase. Finally, syntactically inactive class memberships of the
kind observed in Hiaki stem forms are defined over Vocabulary Items, not over
abstract List 1 morphemes. Indeed, it seems reasonable to think that class
membership is not a diacritic on a Vocabulary Item, but instead is just a reflec-
tion of the presence of that Vocabulary Item on a particular readjustment rule's
list of triggers. This opens the door to the possibility that certain apparently
irregular stem formations simply reflect the presence of a single root on the
lists of two or more readjustment rules. The application of both rules, rather
than one or the other, can produce relatively uncommon bound form~free form
relationships. One case discussed above that might admit of such an analysis
is *bwa'e* 'eat', whose bound form, *bwa'a-*, could result from the interaction
of Truncation and Echo Vowel. It remains to be seen, however, whether such

an analysis is viable for most or all of the irregular bound-stem types. We leave such investigation for future work.

Notes

1. It is also worth noting that derivational suffixes can iterate, and that many of them have bound-stem forms as well (analogous to what Bermúdez-Ortero 2013 calls "suffix stumps" in Spanish). For example, the suffix *taite* 'start' has a bound-stem form *-taiti-*, which occurs when another suffix from the list in (2a) follows it.

2. Siddiqi (2009) proposes an approach that brings a stem-listing analysis of English irregularity into a Distributed Morphology model; the remarks of Embick and Halle (2005), and the arguments advanced here, also pose a challenge to such versions of DM.

3. However, see Bermúdez-Ortero 2013 for a thorough discussion and defense of stem-listing approaches to class properties in the grammar of Spanish; lexical redundancy correspondences over stored forms are argued to account for the systematicity of class-based effects in such a model.

4. Note that *mamam* 'hand' is mandatorily plural when free in the language; it is a member of a substantial class of such *pluralia tantum* nouns in Hiaki, analogous to *scissors* or *pants* in English. It is nonetheless clear that the free-stem form is *mama-*, not *mam-*; when the plural suffix *-m* is added to a consonant-final stem, the epenthetic vowel that repairs the coda cluster violation is *-i-*, not *-a-* (consider *tekil* 'job', *tekilim* 'jobs').

5. Recall that vowel shortening, as in the first syllable of *chiiva~chiva'a-* here, is a regular phonological process in the language that occurs regardless of whether the affixation that triggers it is inflectional or derivational. Hence we do not include "vowel shortening" in our description of the process that forms the bound stem here, since we cannot distinguish the regular vowel-shortening process triggered by affixation from vowel shortening for stem formation.

6. The question of conditioning *across* intervening morphology is of course highly pertinent in other frameworks as well; see, for example, Carstairs 1987 and subsequent work. Since in simple nouns and verbs the intervening morphology at issue here is null, however, the locality of the context under discussion is an issue mostly for theories that make robust use of zero morphs, such as Distributed Morphology. See Embick 2010 for an in-depth discussion of these issues.

7. However, see Bobaljik and Wurmbrand (chapter 11, this volume) for a possible mechanism for suppressing the implementation of cyclic spell-out that relates cyclic behavior to the selectional relationship between the cyclic head and its immediately c-commanding head. It remains to be seen whether such a mechanism could be relevant to the derivation of the Hiaki forms under discussion here.

8. Ora Matushansky (personal communication) asks whether Subclass iii, involving free forms ending in *-u* and bound forms ending in *-oe*, might be derived by a readjustment rule involving raising of the final *-e* to *-i*, followed by assimilation of *-o* to the height of *-i* (producing *-u*) and subsequent deletion of *-i* in the free form. Such an

approach to this subclass may be motivated by the fact that final V-deletion seems to be in evidence in other corners of the grammar, such as in alternate lexicalizations of adverbials like *ket~kechia* 'still'. Nonetheless, final *-ui* VV sequences seem to be well formed in the language, in words like *kumui* 'ego's mother's older brother', so such deletion is not a general phonological process.

9. A large number of verbs in Hiaki have transitive forms that end in *-a* (and whose bound forms are Class 3, invariant) and intransitive ones that end in *-e* (and whose bound forms are Subclass i, *e→i*). The verbs in table 7.2, however, are not among these transitive/intransitive alternating pairs, and contain transitives whose free forms end in *-e* (e.g., *bwa'e* 'eat.TR') as well as intransitives whose free forms end in *-a* (e.g., *hovoa* 'get full').

10. Indeed, we remain agnostic concerning the character of the phonological component that interprets the output of the morphological component. It might well be, as proposed in Haugen 2008, an optimality-theoretic constraint system. In that case, readjustment rules could be analyzed as subphonologies or cophonologies constrained to apply to small subsets of the lexicon. We leave this possibility for future research.

11. A bimoraic requirement on the foot on the right edge of a bound stem would provide a prosodic characterization of the motivation for Truncation, as well, since the result of Truncation seems always to be a heavy syllable.

12. As noted by Matushansky (personal communication), the possibility that certain forms may be the product of the application of two independently active readjustment rules is in itself an argument *for* the existence of readjustment rules and a strike against the listedness hypothesis, in which such interactions could not be captured. Alternatively, one might consider the notion that readjustment rules should be treated as cophonologies involving a few reranked constraints, restricted to a small list of individual lexical items, within an interpretive OT approach to the output of the DM morphological component. In that case, the apparent "dual" application of readjustment rules in the case of forms like *bwa'e~bwa'a* would involve the optimal satisfaction of two lexically restricted cophonologies (see note 10 above). We leave the investigation of such a possibility for future work.

13. Note that a suppletion analysis of stem-form alternations could equally appeal to this context—that is, a rule stating that a bound-stem form is inserted whenever a root node is *not* left-adjacent to a phase edge, or, conversely, that a free form is inserted whenever a root node is left-adjacent to a phase edge, assuming that both syntactic structure and linear order are accessible to the Vocabulary Insertion operation (rather than assuming, for instance, that linear order is determined by the outcome of Vocabulary Insertion, as assumed in, e.g., Harley 2010). This weakens, somewhat, the locality-based objection to the 'listed' approach to stem-form alternations, but does not affect the other arguments presented above. Thanks to a reviewer for this observation.

14. Although the representation here suggests that these nodes are disambiguated by their meaning, in fact we assume they are individuated only by a contentless index, following Harley 2011. The index is interpreted by the equivalent of Vocabulary Insertion at LF—a kind of meaning-lookup procedure.

15. Note that the singular form is the marked variant in Hiaki; in impersonal constructions, where number is unspecified, the plural form surfaces. Also note that the

conditioning environment in (10) represents a syntactic structure, not a purely linear structure; it conveys the notion that the singular form of a root is inserted when that root is sister to a singular DP—that is, when that root takes a singular internal argument. As noted above, Harley, Tubino Blanco, and Haugen (2009) argue that all intransitive suppletive verbs in Hiaki are unaccusative. Other material may come to linearly intervene between the root and its base-generated sister DP, for example by virtue of DP movement to subject position, or the prefixation of an adverbial particle to the verbal complex, but since this Vocabulary Insertion rule is sensitive to syntactic context, not (necessarily) morphological adjacency, and since the rule presumably applies on the first *spell-out* cycle, prior to (e.g.) DP movement, such intervening material will not block insertion of the conditioned verb form.

16. A reviewer notes that one might consider whether the singular and plural forms of these suppletive verbs should not simply be considered independent verbs (i.e., independent List 1 elements), as proposed in Mithun 1988. In such an approach, each "suppletive" pair is made up of two independent verbs that have effectively the same Encyclopedic content but one of which subcategorizes for singular arguments, the other for plural ones. If that were the case, the class features could be attached to the abstract verb rather than to the phonological exponent. The Hiaki cases, all of which involve verbs that subcategorize for an internal argument, do not provide the basis for a decisive argument against this position, but such an analysis would face the (in our opinion) significant drawback of failing to capture grammatically the intuition of Hiaki speakers that the singular and plural forms of these elements are the same verb—that is, that they form part of a paradigm. See Bonet and Harbour 2012 for a more decisive reply to Mithun's proposal based on the behavior of suppletive nominal roots, lacking a selected complement.

8   "Not Plus" Isn't "Not There": Bivalence in Person, Number, and Gender

Daniel Harbour

Halle (1957) gives arguments for adopting binary features in phonetics and phonology, but this is unusual; the internal structure of features tends to be assumed rather than argued for.
—Corbett (2010, 18)

8.1   Introduction

When he was young, Morris Halle taught himself to write backwards (as did I). His method involved decomposing letters into their constituent ascenders, descenders, loops, and humps, and mastering reversal just of this smaller set of primitives. In consequence, an examination of features, the primitives of linguistic representations, is a fitting tribute both to Morris's long and ongoing contributions to linguistics and to the personality evident in his earliest intellectual excursions.

Fundamental to features is the often-neglected issue of valence (cf, Halle 1957 on phonology). Can features be explicitly denied and asserted, or is only assertion explicit, with negation implied by absence? In other words, bivalence or privativity: +F versus −F, or plain F versus nothing?

This chapter aims to review the accumulating case for bivalence of person, number, and gender features, from semantics, morphology, and syntax. The evidence strongly suggests, I believe, that explicit reference to negation is necessary and that negation and absence are distinct.

Since privativity is more restrictive, the initial burden of proof lies on advocates of bivalence. Yet, as Noyer's landmark (1992) study of person and number features assumed bivalence, it compels advocates of privativity to produce equal empirical coverage. This has been attempted to good measure (Harley 1994; Harley and Ritter 2002), with Harley 1994 even addressing one case, Mam, where Noyer explicitly argued for bivalent person features.

However, Noyer's case for the bivalence of number, from Kiowa-Tanoan, has not been reanalyzed in privative terms and Harbour 2011d argues that it cannot be. Moreover, arguments against the privativity of (some) person features occur in Nevins 2007 (on the typology of person-case constraints), Trommer 2008 (on Dumi and Menominee), and Watanabe 2012 (on affix order in Fula).

Because my view of person (2012) differs substantially from Trommer's, Nevins's, and Watanabe's, I do not review their arguments here (though I agree with much of them). Instead, I present evidence from four sources, some of which have received little theoretical attention. Section 8.2 examines the semantic composition of trial, unit augmented, and greater paucal in terms of opposing specifications, $(+F(-F(...)))$, of a single feature. Section 8.3 treats the morphological composition of persons and numbers that show complementary patterns of inclusion (for instance, inclusive, which sometimes includes exclusive, but is sometimes included by it). Section 8.4, the most traditional, presents "alpha exponents," which track covariant values of multiple features. And section 8.5 highlights the role that bivalence plays in one theory of the strong person-case constraint (Adger and Harbour 2007) and shows that this permits the account to extend to previously unnoted phenomena in differential marking of (in)animate objects in Tewa.[1]

8.2 Semantic Compositionality: Function Application of +F to −F

A common aim in morphosemantics is to derive maximal coverage from minimal posits. A major result along these lines is the discovery that the (two) features that distinguish singular from minimal pronouns suffice to derive the six number values, comprising three distinct number systems (singular–plural, minimal–augmented, singular–dual–plural), with the appropriate differences in lower bounds of the two plurals (two in dualless languages, three otherwise) (Noyer 1992; see Harbour 2011a for further exposition). In this light, a semantic argument for bivalence comes from the feature composition of trial, unit augmented, and greater paucal, numbers unanalyzed on most accounts. Trial and unit augmented require no features beyond those just alluded to, ±atomic and ±minimal.[2] And greater paucal needs no feature beyond ±bounded, which characterizes the basic paucal itself. Rather, these previously recalcitrant numbers arise from function application of +F to −F. Thus, a putative feature ±trial is as redundant as ±dual, provided our basic features are bivalent, because only bivalent features afford the −F for +F to apply to.

Consider how, featurally, one would add trial to the system singular–dual–plural. For any person or noun n, $(-at(n))$ picks out the nonatoms of n. The most minimal nonatoms are dyads; hence $(+min(-at(n)))$ is the dual. The nonminimal remainder, $(-\text{min}(-\text{at}(n)))$, is the plural. Many languages end their number systems there. However, it is possible, once again, to pick out the most minimal elements of the plural, $(+\text{min}(-\text{min}(-\text{at}(n))))$, which yields the trial.

Crucially, this result depends on distinguishing +min and −min both from absence and from each other. If minus were absence, then dual [+min −at] and trial [+min −min −at] collapse to just [+min]. If plus were absence, trial would collapse to plural [−min −at].

Minimal, unit augmented, and augmented (e.g., Bininj-Gunwok, Evans 2003) correspond to dual, trial, and greater for first-person inclusive, but to singular, dual, and greater for all other persons and nouns. The system arises from ±min alone. The smallest inclusive, $(+\text{min}(1\text{IN}))$, is the speaker-hearer dyad. So, $(-\text{min}(1\text{IN}))$ is the speaker, hearer, and one or more others. The smallest of these, $(+\text{min}(-\text{min}(1\text{IN})))$, is the speaker, hearer, and just one other—that is, the unit augmented. For first-person exclusive (other persons and nouns are analogous), $(+\text{min}(1\text{EX}))$ is the speaker alone; $(-\text{min}(1\text{EX}))$ is the speaker and one or more others; and $(+\text{min}(-\text{min}(1\text{EX})))$ is the smallest of these, the speaker and just one other, the unit augmented.

One can define the paucal in terms of +bounded owing to the observation that a plurality plus a plurality is always a plurality, but a paucity plus a paucity is not always a paucity. That is, plurals are unbounded under addition, but paucals are bounded. So, in, say, Yimas (singular–dual–paucal–plural), paucal is $(+\text{bdd}(-\text{min}(-\text{at}(n))))$ and plural, $(-\text{bdd}(-\text{min}(-\text{at}(n))))$. A further option that a language may exploit is to carve out, from the latter unbounded region, a second bounded subregion, $(+\text{bdd}(-\text{bdd}(-\text{min}(-\text{at}(n)))))$. This picks out elements larger than paucal, but still of limited size. The Sursurunga system (singular–dual–paucal–greater paucal–plural) results.

It is a considerable economy to reduce recalcitrant numbers to features that are independently required. Yet, the observation made for the trial applies also to the unit augmented and greater paucal: the necessary distinctions are only possible where +F and −F are both represented and both distinct from zero. This provides a semantic argument for bivalence.

Two consequences of this approach are worth nothing, in distinction to Harley and Ritter 2002. The latter assumes that trial and paucal are different interpretations of the same feature structure. Yet, if paucals permit "exact" interpretations, then the exact counterpart of the greater paucal should be the quadral. Typological surveys suggest no quadrals exist (Corbett 2000; Lynch, Ross, and Crowley 2002). The current approach excludes the quadral, on the assumption that features come in sets. Tetrads are the minimal (+min) elements, once dyads $(-\text{min}_1)$ and triads $(-\text{min}_2)$ have been excluded from the nonatoms, $(+\text{min}(-\text{min}_2(-\text{min}_1(-\text{at}(n)))))$. Yet, as a set, [+min −min −min −at] is indistinguishable from [+min −min −at] (by the axiom of extension, $\{a, a\}$ = $\{a\}$). Thus, axioms of set theory prevent one from building the set that would denote the quadral: if one adds an extra −min to the trial, the axiom of extension "gobbles" it up.

Moreover, if, as per Harley and Ritter 2002, trial and paucal are different interpretations of the same structure, then they are predicted not to co-occur. The current approach makes the opposite prediction: trial and paucal are featurally distinct and so may co-occur, just as dual and paucal do in Yimas and like systems. The latter is the correct position, because trial and paucal co-occur in Lihir, Marshallese, and Mussau (Corbett 2000; Ross 2002; Brownie and Brownie 2007).

Thus, distinguishing +F and −F from each other and from their absence permits one to characterize trial, unit augmented, and greater paucal in a uniform and semantically orthodox fashion (by function application), while deriving significant facts about nonexistent and co-occurrent numbers.[3]

8.3 Morphological Compositionality: Reciprocal Inclusion

By "morphological compositionality," I intend phenomena like the famous Hopi dual (Hale 1997), which shares one exponent (a suppletive verb) with singular and another (a pronoun) with plural. This has been taken to argue for analyses in which dual shares features with both singular and plural. The Hopi facts are shown in table 8.1, with privative (Harley and Ritter 2002) and bivalent analyses; underlining shows shared exponents and the features they realize. Analogous phenomena arise for person, in Tok Pisin pronouns (Foley 1986; ignoring number, inclusive *yumi* is second *yu* plus first *mi*), and in Kiowa (Harbour 2007; table 8.2; IN/EX share a pronoun, IN/2, agreement).[4]

Table 8.1
Hopi: A composed dual

	'I/we'	'ran'	Privative	Bivalent
SG	*nu*	*wari*	min	+min(+at)
DL	*'itam*	*wari*	min grp	+min(−at)
PL	*'itam*	*yu'tu*	grp	−min(−at)

Table 8.2
Kiowa: A composed inclusive

	PRN	AGR-be	Privative	Bivalent
1EX	*náw*	*e-dáw*	spkr	+spkr(−hear)
1IN	*náw*	*ba-dáw*	spkr hear	+spkr(+hear)
2	*ám*	*ba-dáw*	hear	−spkr(+hear)

Although privative and bivalent analyses are equally matched for Hopi and Kiowa, they come apart when relationships of compositionality are reversed. In their respective pronouns, the Damana dual is a component of the plural (Amaya 1999; table 8.4 shows second person) and the Limbu inclusive is a component of the exclusive (Van Driem 1987; table 8.4 shows the dual).

Table 8.3
Damana: A composed plural

PSN-NMB-NMB	Privative	Bivalent
SG *ma*	min	+min(+at)
DL *ma-bi*	min grp	+min(−at)
PL *ma-bi-nyina*	grp	−min(−at)

Table 8.4
Limbu: A composed exclusive

PSN-DL-PSN	Privative	Bivalent
1EX *ang-chi-ge*	spkr ___	+spkr(−hear)
1IN *ang-chi*	spkr hear	+spkr(+hear)
2 *khen-chi*	hear	−spkr(+hear)

A bivalent analysis is trivial in both cases, given the availability of negative features: Damana *nyina* is −min, Limbu *ge* is −hearer. In the privative systems of tables 8.3 and 8.4, however, these morphemes emerge like phantoms from the void: there is no privative feature corresponding to Damana *nyina* or Limbu *ge*. These problems are not restricted to second person in Damana (first- and third-person plural also "add" *nyina* to dual PSN-*bi*), nor to duals in Limbu (where verb paradigms also show "addition" of *ge* to form exclusives from inclusives).

Nor are these phenomena isolated to the languages just mentioned. Comparable to Damana are, for instance, Mokilese (Hutchisson 1986; e.g., 2DL *kamwa* is a component of 2PL *kamwa-i* and 2GR.PL *kamwa-ⁱ* = *kimwi*) and Walapai (Redden 1966; e.g., 2PC *máč* is a component of 2PL *máč-uv*). And comparable to Limbu is Nyawaygi (Dixon 1983; e.g., 1IN.DL *ngali* is a component of 1EX.DL *ngali-lingu*). However, so far as I know, advocates of privative features have yet to address systems such as these, and my attempts to sketch out a phantom-free analysis with privative features suggest deficiencies within privativity itself.

Consider Limbu. One could posit a new feature [exclusive], with exponent *ge*.[5] But, as [exclusive] means 'excludes speaker', this simply shows, contra privativity, that one does need to refer to what negative values refer to. Nor are matters helped by, say, taking *ge* to be an exponent of number. One still needs a means of saying that *ge* occurs in the absence of [hearer]. Again, the need to "refer" to negative values remains.

Alternatively, if one adds a feature, [person], to all persons, then the analysis in table 8.5 is made available (cf, Harley and Ritter's [PARTICIPANT]). The table reads left to right as a derivation; column headings are exponents that apply successively. The table delivers the desired result, but it is somewhat ad hoc: without contextualization (/spkr), *ang* would occur in second person 2; without

Table 8.5
Limbu: A privative treatment, phantom-free

	Input	pers$_{/spkr}$ ⇔ *ang*	spkr hear ⇔ Ø	hear ⇔ *khen*	spkr ⇔ *ge*	Output
1EX	pers spkr	*ang* spkr	*ang* spkr	*ang* spkr	*ang* ge	*ang* . . . *ge*
1IN	pers spkr hear	*ang* spkr hear	*ang* Ø	*ang* Ø	*ang* Ø	*ang*
2	pers hear	pers hear	pers hear	pers *khen*	*khen*	*khen*

the zero exponent, *khen* and *ge* would occur in 1IN; and stipulation of order is crucial (Ø before *ang* would block *ang* from 1IN).[6] To claim that systems like Limbu arise only owing to ad hoc vocabulary specifications is to afford them a very different status from Hopi number, which has been central to morphosemantic thought.

Privative features can easily deal with morphological compositionality where "more features" corresponds to "more exponents." The reverse situation is problematic, but is attested in a range of person and number systems. These show that, where two bivalent features suffice, two privative features do not. Given that negation or other forms of complementation must be available to the semantics (*not* and *the rest*, after all, have interpretations), it is a minimal increase in complexity to represent affirmation/negation as part of the features. If better privative analyses of such systems are forthcoming, this will hardly detract from the naturalness with which bivalent features can capture their properties.

8.4   Covariant Values (Alpha Exponents)

The preceding argument used different exponents to show that both plus and minus are represented and realized. Sometimes, a single exponent suffices to show this, if it realizes both [+F +G] and [−F −G], or both [+F −G] and [−F +G]. I term these "alpha exponents," because they have the form [αF αG] or [αF −αG], where, as in Chomsky and Halle 1968, α is a variable over + and −.

A well-known proposal of an alpha exponent arises in Noyer's (1992) treatment of Mam. In possessives and ergatives, enclitic *a* occurs for the seemingly unnatural class of 1EX and 2 (England 1983). Noyer observes that this corresponds to αspeaker −αhearer.

Alpha exponents make an excellent case for bivalence. Yet Mam *a* concerns a single morpheme in a single language and has been subject to privative reanalysis.[7] However, there are at least three domains where alpha exponents are more widespread, both intra- and interlinguistically: epistemic alignment, number-dependent gender, and inverse number systems (see also Béjar and Hall 1999; Baerman 2007; Albright and Fuß 2012; Wunderlich 2012).

Like Mam *a*, epistemic alignment involves alpha exponence for person. The core characteristic of the phenomenon is that one and the same set of suffixes marks first person in assertions and second person in questions, whereas a different set marks first person in questions and second person in assertions (as well as third person in both). In Awa Pit (Curnow 2002), for instance:

(1) na-na   pala   ku-mtu-s̲   min-ta-ma ashap-tu-y̲?
    1SG-TOP plantain eat-IMPF-<u>ALIGN</u>   who-ACC-Q annoy-IMPF-<u>NONALIGN</u>

    'I am eating plantains.'   'Whom am I annoying?'

(2) nu-na   pala   ku-mtu-y̲   shi-ma ki-mtu-s̲?
    2SG-TOP plantain eat-IMPF-<u>NONALIGN</u> what-Q do-IMPF-<u>ALIGN</u>

    'You are eating plantains.'   'What are you doing?'

(3) us-na   atal   ayna-mtu-y̲   mintas   a-mtu-y̲?
    3SG-TOP chicken cook-IMPF-<u>NONALIGN</u> where from come-IMPF-<u>NONALIGN</u>

    'He/she is cooking chicken.'   'Where is he coming from?'

Different languages vary as to specific details (such as the treatment of nonagent first persons and of third-person logophors in subordinate clauses; see Curnow 2000 for comparison of Kathmandu Newari and Lhasa Tibetan (Tibeto-Burman), and Tsafiki and Awa Pit (Barbacoan)). But confining attention to the core sensitivity illustrated above, we have use of one exponent, *s*, for (non–third person) αauthor −αQ, and use of another, *y*, otherwise.

In a different vein, some languages make gender dependent on number via alpha exponence (or, perhaps, via alpha rules that swap plus and minus prior to exponence). In Hebrew, for instance, "The cardinal numbers from 3 to 10 [have] this peculiarity, that numerals connected with a masculine substantive take the feminine form, and those with a feminine substantive take the masculine form" (Kautzsch 1910, §97*a*; the same holds true for Modern Hebrew). For instance, *šaxor* and *šxor-a* are 'black', masculine and feminine, but *šaloš* and *šloš-a* are 'three', feminine and masculine. That is, these numerals have gender −αfeminine when modifying nouns of gender αfeminine.[8] Somewhat similar are nouns like 'egg' and 'arm' in Italian (and 'sense' in Romanian), which are masculine in the singular but feminine in the plural. For example, *il bianco*, 'the white one', could refer to an egg or a man but not to a woman, but *le bianche*, 'the white ones', to eggs or women, but not to men (see Acquaviva 2008). Nouns of this class have gender −αfeminine for number αatomic.

But doubtless the most striking case of alpha exponence arises in the "inverse" number systems of the Kiowa-Tanoan languages. As originally argued in Noyer 1992, these languages define noun classes via number features. That is, where Hebrew and Italian nouns are inherently ±feminine, Kiowa-Tanoan has nouns that are inherently ±atomic and/or ±minimal.[9] Table 8.6 illustrates part of the Kiowa system. There are no singular or plural markers per se. Rather, the "inverse" (*dau* for the nouns chosen here) marks singular for 'stick', plural for 'bug', singular and plural for 'tomato', and is absent from 'rock'. Assuming the noun-class features shown beside each noun in the table, we can characterize the distribution of *dau* by claiming that it realizes [αF −αF], where F is either number feature. The conflicting features are shown in square brackets after each noun, with the feature contributed by noun class on top, that contributed by number below. For example, PL('bug') contains [+min −min] and so is inverse-marked *pól-dau*, because 'bug' is +min but plural is −min. Similarly, SG('stick') is inverse-marked because it contains [+at −at], −at from 'stick', +at from singular. Both these conditions apply to 'tomato', but 'rock', which has no class features, does not ever clash with number and so is never inverse-marked. See Noyer 1992 and Harbour 2007, 2011d, for further discussion and application to other members of the family.

Whether privativity can handle epistemic alignment and number-dependent gender remains to be seen (Béjar and Hall 1999 use markedness metrics). However, as mentioned in the introduction, Harbour 2011d argues that privative features cannot insightfully characterize the inverse systems of Kiowa and Jemez (the obvious, but far from only, problem being that the notion of value conflict, which is the essence of the inverse, cannot readily be expressed in a system where features are valueless). By contrast, bivalent features readily accommodate alpha exponence. So, on balance, the phenomenon supports bivalence of person, number, and gender features.

Table 8.6
Kiowa inverse marking

	'bug' (+min)	'stick' (−at)	'tomato' (+min −at)	'rock' (Ø)
SG (+min +at)	*pól*	*áa-dau* $\left[\begin{smallmatrix}-\text{at}\\+\text{at}\end{smallmatrix}\right]$	*k'âun-dau* $\left[\begin{smallmatrix}-\text{at}\\+\text{at}\end{smallmatrix}\right]$	*ts'ów*
DL (+min −at)	*pól*	*áa*	*k'âun*	*ts'ów*
PL (−min −at)	*pól-dau* $\left[\begin{smallmatrix}+\text{min}\\-\text{min}\end{smallmatrix}\right]$	*áa*	*k'âun-dau* $\left[\begin{smallmatrix}+\text{min}\\-\text{min}\end{smallmatrix}\right]$	*ts'ów*

8.5 Person Sensitivities

The final set of arguments I will put forward are morphosyntactic, building on the account, in Adger and Harbour 2007, of the strong person-case constraint—the requirement, in some languages, that, when agreement or clitics represent both indirect and direct object, the direct object must be third person. Section 8.5.1 reviews the account, emphasizing the role that bivalence of ±participant plays, and section 8.5.2 extends it to novel data from Tewa indirect and animate direct objects. For reasons of space, I do not attempt a privative recasting of the account. However, the challenge to privative views of person features will be clear: a three-way contrast between plus, minus, and absence is crucial, and only bivalent features afford this.

8.5.1 Person and Case: Constraint and Syncretism

The central motivating observation of Adger and Harbour 2007 is that, in many languages with the strong person-case constraint, first- and second-person agreement or clitics display a syncretic behavior from which third person is exempt: only for third persons are indirect-object (dative) and direct-object (accusative) forms distinct. In French, for instance, 1SG *me*, 2SG *te*, 1PL *nous*, and 2PL *vous* are the clitics for both indirect and direct object; for third person, 3M.SG *le*, 3F.SG *la*, and 3PL *les* are used for direct objects, but 3SG *lui* and 3PL *leur*, for indirect objects. A similar pattern holds across Romance and Kiowa-Tanoan, as well as in Chinook, Georgian, and Yimas. In other words, the persons prohibited from object position in PCC configurations are those that (generally) display this syncretism.

Adger and Harbour 2007 accounted for this correlation by attributing two roles to ±participant. First, it is an indispensable part of the semantics of first and second person. Whereas third person can be specified solely for number (and gender), other persons must also be specified for ±participant (inter alia). Second, ±participant encodes the well-known animacy constraint on many applicatives (e.g., Perlmutter 1971). We claim that syntax does not have a feature ±animate per se and must represent animacy with the next best thing; ±participant is a good fit because it is an essential ingredient of the indisputably animate first and second persons. So, Appl, the head that selects the applicative argument, does so by agreeing with it for ±participant.

The syncretism arises for first and second person because the selectional requirement on applicatives makes no difference to their feature content. They are specified for ±participant whether selected by Appl or not. This is not so for third persons: selection by Appl entails a −participant specification that they would otherwise lack, and this is something to which exponence can be sensitive. For the French clitics above, for instance, *le* and *la* are plain

[±feminine +atomic], but *lui* is [−participant ±feminine +atomic]; and *les* is plain [±feminine −atomic], but *leur* is [−participant ±feminine −atomic]. The difference between dative and accusative for third person, then, is the contrast between minus and absence of ±participant.[10]

The strong person-case constraint arises because Appl case licenses the lower argument before it licenses its specifier. Both case licensing and selection involve agreement in phi-features, and the uninterpretable ±participant feature of Appl can agree (be valued) only once. If the lower argument is specified for ±participant, then Appl has no resource left by which to license its specifier. Consequently, an applicative argument can occur only if the lower argument has not "used up" Appl's ±participant feature. That is, the direct object must lack a ±participant specification, and so must be third person.

This account of the person-case constraint and its concomitant syncretisms makes full use of the resources of bivalence. It distinguishes between +participant and −participant in the characterization of agreement and clitics for first- and second-person versus third-person indirect objects. And it distinguishes between minus and absence in the characterization of agreement and clitics for third-person indirect versus direct object. Insofar as one accepts the account—and new evidence from Tewa suggests one should—one has evidence for bivalence of ±participant.

8.5.2 Animate Direct Objects in Tewa

Tewa, a Tanoan language related to Kiowa, provides strong, and previously unnoted, evidence for the necessity of distinguishing between third persons specified as −participant and those unspecified for ±participant. Where Kiowa (like the other languages examined in Adger and Harbour 2007) uses −participant only for third-person applicatives, Tewa uses it for some direct objects too. As with applicatives, this induces an animacy effect. So, in Tewa, animate direct objects may be differentiated from inanimate direct objects by being specified as −participant. The extension of dativelike properties to animate direct objects paints Tewa as *leísta* Kiowa (Ormazabal and Romero 2007), a parallel that some readers may find helpful. This ramifies in four quite diverse ways throughout Tewa grammar.[11]

8.5.2.1 *Agentive Marking*

Agents in Tewa are (optionally) marked with -*d̲i*:

(4) I   pu'ay-d̲i   páad̲é-bo óe-mû'   i   P'osewhâa Sedó
     the rabbit.DIM-<u>AGT</u> first-PRT 3:3SG.AN-see.PF the coyote   old man

     'The little rabbit saw Old Man Coyote first.'

(5) I   pu'ay   óe-yóe'an   i P'osewhâa Sedó-d̲i
    the rabbit.DIM 3:3SG.AN-leave.PF the coyote   old man-AGT

    'Old Man Coyote left the little rabbit.'

(6) Naa-d̲i wí   píví   wîn-hóewayní
    I-AGT   some meat 1:2SG:3-go get.FUT

    'I'll go get you some meat.'

(7) Hed̲i i k'úwá'ay-d̲i in   p'ônbay ôn-'íh-kê'
    and   the lamb-AGT   the.INV head 3:3SG:3-back-take.PF

    'So the lamb hoisted the skull onto his back [for the kitten].'

However, this is only possible for agents that act on animate direct objects
(4)–(5) and/or on indirect objects (6)–(7). In no example in the *Téwa Pehtsiye*
corpus is an agent marked with -*d̲i* when its only clausemate argument is
inanimate. The following are typical of such scenarios:

(8) Hed̲i-ho i-kê'   i   tųųyó in   yán   t'ún
    and-PRT 3SG:3-get.PF the chief the.INV willow basket

    'The chief took the willow basket.'

(9) Wí Kwą́k'u kwiyó   i-kaysu-kunmáa
    one Spanish old woman 3SG:3INAN-cheese-sell.IMPF

    'An old Spanish woman was selling cheese.'

This can be explained simply. Indirect objects and animate direct objects
form a natural class that excludes inanimate direct objects, by virtue of specification for ±participant. And the dual role of v, in case licensing the next
argument down and in selecting the agent, makes sense of how the agent comes
to "register" the feature content of a lower argument. We can regard -*d̲i* as an
exponent of v on the agent, when v has agreed with a lower ±participant as a
part of case licensing.

8.5.2.2 Incorporation

Inanimate direct objects in Tewa may incorporate, as in (9)–(11).

(10) Naa-d̲i wây-píví-má'í
     I-AGT   1:2PL:3-meat-bring.FUT

     'I'll bring you the meat.'

(11) Thamuwaagá . . . i-'ągą̈h-khą́ą̈-k'o'
     every day.FOC   3SG:3-atole-scum-eat-IMPF

     'Every day he ate the scum from the atole.'

The animate direct objects and indirect objects in our corpus never incorporate, by contrast. For indirect objects, this restriction may be structural (e.g., Baker 1988). For direct objects, however, an additional factor must be at play. Given that first- and second-person pronouns also fail to incorporate—(12) and (13) are typical of their distribution—it is plausible that ±participant, or the structure that hosts it, prevents incorporation: objects specified for the feature (first and second person, and third-person animates) cannot incorporate, others (inanimates) can.[12]

(12) To'dan naa dí-k'owa?
    who   I   3:1-cut hair.PF

    'Who cut my hair?'

(13) Naa dí-k'oe-d̲a . . .
    I   2:1-eat-REL.FOC

    'If you eat me . . .'

8.5.2.3 Syncretism

The extended use of −participant induces new systematic syncretisms, between indirect and animate direct third person objects. This arises because, for any given number, the two are featurally identical. So, whether v agrees with (case licenses) 3SG, say, as the indirect object of a ditransitive or as the animate direct object of a monotransitive, it will be valued in the same way, [−participant +minimal +atomic], yielding identical agreement morphology. Table 8.7 shows this in full, using the sentence frames '*x* verb it to *y*' for ditransitives and '*x* verb *y*' for transitives.

The table shows that animate direct objects pattern with indirect objects in two ways. First, these prefixes are invariant for the number of the agent.[13] For example, *óe* encodes any third person acting on third-singular animate—that is, action on 3SG.AN by a 3SG, 3DL, or 3PL agent. Similarly, *ôn* encodes giving

Table 8.7
Tewa agreement prefixes for third-person argument combinations

x		*x* VERB it to *y*			*x* VERB *y*.AN			*x* VERB *y*.INAN		
		3SG	3DL	3PL	3SG	3DL	3PL	3SG	3DL	3PL
y	3SG	*ôn-*			*óe-*			*i-*	*dân-*	*dây-*
	3DL	*ovân-*			*ovân-*					
	3PL	*ovây-*			*ovây-*					

to 3SG by a 3SG, 3DL, or 3PL agent. The sensitivities are reversed for inanimate objects, however: only agent, not object, number is relevant. For instance, *i* encodes third singular acting on any third person—that is, 3SG acting on a 3SG, 3DL, or 3PL inanimate.

Furthermore, the prefixes for indirect objects and animate direct objects are made up of the same exponents. Indeed, for 3DL and 3PL, the prefixes are identical. *Ováy*, for instance, is used for any third person either giving something to 3PL, or acting transitively on an animate 3PL. Identity all but obtains for 3SG too, where the animate transitive prefix is *óe* /ó:/ and the ditransitive is *ôn*. Other prefixes show the same alternation, witness *dí* and *dîn* (as in 'you VERB me' and 'you VERB it to me'), *wí* and *wîn* ('I VERB you' and 'I VERB it to you'), and *wóe* /wó:/ and *wôn* ('he VERBS you' and 'he VERBS it to you'). Thus, *-ˆn* expresses inanimate-object agreement in ditransitives. Subtracting this leaves identical *óe* for both −participant 3SG forms.[14]

These systematic correlations thus strongly support an account that regards third-person indirect and animate direct objects as featurally identical.

8.5.2.4 Animacy Restrictions

I also predict that animate direct objects cannot bear −participant in a ditransitive. If Appl case-licensed such an object, its participant feature would be valued and it could no longer select a specifier. This is the configuration of the person-case constraint. To avoid it, animate objects must be featurally represented as inanimates, when there is an applicative.

Verifying this prediction requires attending to the composition of the agreement prefixes themselves (an area of notorious complexity across the family; see, e.g., Watkins 1984 or Harbour 2007 on Kiowa). A comparatively straightforward illustration is provided by the following triplet:

(14) Năwe dovân-'ąh-khęh-hon
       here   1:3DL.AN-foot-chase-bring.PF

       'This is where I've tracked them to.'

(15) Dîn-pee-yôn
       2:1:3-exit-command.IMP

       'Tell them [DL] to come out [for me].'

(16) Dîn-ts'úde-í   in   to   khän p'ônbay
       2:1:3-bring in-FUT the.INV ANAPH lion head

       'Bring me the skull of the lion.'

'Them' in (14) and (15) refers to the same pair, the protagonist kitten and lamb. (14) illustrates that they are treated as animates. Its prefix, *dovân*, decomposes into two readily recognizable parts. There is the first-person subject/object exponent *d-*, present in the *dîn* of (15)–(16) and the *dí* of (17)–(18):

(17) To'dan naa dí-k'owa?
      who   I   3:1-cut hair.PF

      'Who cut my hair?'

(18) Naa dí-k'oe-d̲a . . .
      I   2:1-eat-REL.FOC

      'If you eat me . . .'

Subtracting *d-* from *dovân* leaves *-ovân*. This is recognizable from table 8.7 as the prefix used when acting on a 3DL animate direct object (assuming the third-person agent makes no morphological contribution, as was already shown for agent number). It follows that *dovân* in (14), like *ovân* in table 8.7, registers the animacy of its direct object.

If the 3DL direct object triggered animate agreement in (15), we would expect the prefix to consist of some combination of *dí* for 2:1 (18) and *ovân* for animate 3DL (presumably, *díovân*, *dóvân*, *dívân*, or similar). Instead, the direct object contributes just *-ˆn*, like the inanimate in (16). So, object animacy is indeed suppressed exactly where it would violate the person-case constraint.

8.5.3 Summary

The syncretisms and restrictions on animate agreement, agent marking, and incorporation above follow naturally if −participant distinguishes animate from inanimate direct objects, within the framework of Adger and Harbour 2007. Like that treatment of the person-case constraint and its syncretisms, this crucially distinguishes −participant from its absence, and so requires bivalence.

8.6 Outlook

Given its empirical breadth and theoretical variety, the case for bivalence is compelling. There is, it seems, a considerable burden of proof to be shouldered by advocates of privativity for person, number, and gender features.

Equally, though, advocates of bivalence must pick up the privative gauntlet of Béjar 2003, 2008, Béjar and Rezac 2003 (works that address agreement displacement and related phenomena; on systemic typology and markedness, such as Harley and Ritter 2002, Cowper 2005, see, e.g., Harbour 2011a, 2011c, Nevins 2011). I regret that space limitations preclude discussion of these results here. Morris would console me with Ethics of the Fathers 2:21. I thank him with Proverbs 18:4, and therefore 23:12.

Notes

My thanks to David Adger and an anonymous reviewer for helpful comments, and to two research assistants, Itamar Kastner (section 8.3) and Kyle Helke (section 8.5). This work, and Kastner's contribution, were funded by the Arts and Humanities Research Council of the United Kingdom, grant AH/G019274/1: Project א (Atomic Linguistic Elements of Phi), Subproject ב (Nothingness Underrepresents Negation).

1. Several arguments below turn on the three-way distinction between +F, −F, and absence of ±F. I do not regard absence as a value (the zero value, ØF; cf. Halle and Marantz 1993). Absence of ±F has no semantics, nor, do I believe, can it have exponents. Nonetheless, bivalent features create three-way distinctions when absence is taken into account, just as monovalent (privative) features create two-way distinctions.

2. These are Noyer's ±singular and ±augmented respectively. I rename them for ease of exposition. When there are no enclosing brackets, reference is made to the feature itself: semantic expressions are enclosed in parentheses; syntactic feature bundles, in square brackets.

3. For further formal detail and exploration—including morphological compositionality of the numbers discussed, the treatment of "greater" and "even greater" plurals, the unavailability of "even greater paucal," Greenberg-style implications between numbers, and the basis and acquisition of numeral systems—see Harbour 2011b, 2011c.

4. For simplicity, I use Noyer's features. Harbour (2012) offers a more parsimonious alternative.

5. For clarity, privative features occur between square brackets in the main text.

6. Three other sets of exponents achieve the same result. None is less arbitrary than that in table 8.5, as the reader can easily verify.

7. Harley (1994) analyzes Mam with privative features subject to co-occurrence restrictions (a "geometry"). However, the explanatoriness of geometries has been called into question, both in phonology (Halle 1995) and in morphology (Harbour and Elsholtz 2011).

8. Classical Hebrew presents an even more surprising case. Following "consecutive *and*" (which has a particular narrative, i.e., locutionary, force; Kautzsch 1910, §§111–112), perfectivity is expressed by imperfective exponents, and conversely. For instance (ignoring stress shift), *tāp̄aś-tā* is perfective (seize.PF-2M.SG.PF, 'you seized'), but *w-ṭāp̄ś-tā*, imperfective (and.IMPF-seize."PF"-2M.SG."PF", 'you will seize'); and *ti-ṭpoś* is imperfective (2M.SG.IMPF-seize.IMPF, 'you will seize'), but *wat-ti-ṭpoś*, perfective (and. PF-2M.SG."IMPF"-seize."IMPF", 'you seized'). This 'and' itself expresses aspect (imperfective *w*, perfective *waC*). So, exponents of (or conditioned by) αperfective are used for −αperfective in the context of −αperfective 'and'. Given the featural connection between aspect and number explored in Harbour 2011c (following a substantial tradition that includes Link 1983 and Krifka 1992), this case too might count as evidence for bivalence of number features.

9. I leave aside ±group, which distinguishes (non)collective pluralities. See Harbour 2007, 2011d, for details.

10. This accounts for tendencies, not categorical effects. Syncretism could arise in third persons, if exponence ignores −participant. Nonsyncretism could arise in first or second, if exponence is sensitive to the surrounding context, such as the presence of Appl itself (see Adger and Harbour 2007 on Greek singular clitics, for instance). The proposal thus delineates what the dominant pattern of (non)syncretism should be, but allows the smudginess typical of exponence.

11. Example sentences are from *Téwa Pehtsiye: Tewa Tales, San Juan Dialect* (1982), a native-speaker-produced collection in linguistically accurate orthography, with idiomatic English translations. The glossing is my own and owes much to Kyle Helke, who, for a research practicum, retyped and partly glossed the texts (using Martinez 1983 and my hunches) and checked several generalizations discussed below (Helke 2011). I further verified these against materials in Santa Clara and Arizona Tewa (Harrington 1946; Kroskrity and Healing 1978; Kroskrity 1985, 2010).

12. An obvious alternative is that definiteness blocks incorporation, as nearly all our animate objects, as well as first- and second-person pronouns, appear to be definite. However, if we take the native speakers' translations at face value, then (10) and (11) speak against this. Moreover, in another Tanoan language, Southern Tiwa, definiteness clearly does not impede incorporation, since incorporates may be modified by demonstratives external to the verb, as in 'I saw those men' (*yedi bi-seuan-mũban*, that.INV 1s:3INV-man-see.PF; Allen, Gardiner, and Frantz 1984).

13. In fact, as we will see in the next section, there is evidence that the third-person agents make no morphological contribution to the prefix at all. For example, 'I VERB them it' is *dovây*, the addition of first-person *d-* to *ovây* '(he/they) VERB them it'; and 'they VERB you it' is *wovây*, the addition of second-person *w-* to *ovây*.

14. Tewa, like Kiowa, lacks superheavy codas, so emergence of (*w*)*ôn* from (*w*)*ô:n* is just low-level phonology. For unknown reasons, -ˆ*n* occurs only in some (generally monosyllabic) agreement prefixes: it occurs, for instance, in *bîn* 'you VERB it to them', but not in table 8.7's *ovây* 'they VERB it to them' (which is not *ovây* + ˆ*n* = *ovên*). The patterns of distribution are complex and may represent an arbitrary morphological fact about the language. This potential arbitrariness is orthogonal to the parallels highlighted in the main text.

9 Morphemes and Morphophonological Loci

David Embick

9.1 Introduction

Some questions in linguistics have persisted through hosts of theoretical changes. The conflict between affixless and morpheme-based theories raises questions of this type. In its contemporary incarnation, at least two significant objections raised against affixless theories are that they (i) render the interface between syntax and morphology opaque, and (ii) have serious difficulties with the analysis of blocking (e.g., Halle 1990; Noyer 1992; Marantz 1992; Halle and Marantz 1993; Embick 2000; Embick and Halle 2005; Embick and Marantz 2008). Nevertheless, the tension between morpheme-based and affixless theories is as relevant as ever (see section 9.3). My objective here is to develop a further line of argument in favor of morphemes and against affixless theories, one that also opens up new questions in the study of morphophonology.

Nonaffixal morphological alternations—that is, phonological alternations that are morphologically triggered or targeted—are often taken to provide evidence for affixless theories. In this chapter I develop an argument for the opposite conclusion. The argument is based on the observation that morphophonological changes behave as if they have a *morphological locus*—that is, they operate in a way that is expected if they are linked directly to a morpheme that has a position (hierarchically and linearly) within a complex word, and act in a way that is (phonologically or morphologically) local to that morpheme. This aspect of nonaffixal morphology is a component of a broader theory of morphophonological locality, one based on morphemes and the principles governing their composition into complex objects (see Embick 2010, 2012). Crucially, to the extent that the correct theory of morphophonological loci follows from a morpheme-based theory of morphology, significant generalizations about morphophonology are missed in affixless frameworks.

Ideas along these lines have been advanced in different forms in the literature. In my view, however, these points have neither been fully appreciated, nor developed in sufficient detail. I first outline a generalized theory of morphophonological loci in section 9.2, and then illustrate difficulties for affixless theories in section 9.3; section 9.4 concludes.

9.2 A Morpheme-Based Theory of Loci

The empirical focus of this chapter is on different types of evidently nonaffixal alternations, of the types illustrated in (1)–(3).[1] German umlaut is vowel fronting triggered by several morphemes that have nothing in common, as far as the synchronic grammar is concerned (see, e.g., Lieber 1980, 1987; Kiparsky 1996; Wiese 1996a, 1996b; Embick and Halle 2005):

(1) | *Basic* | *Umlauted* | *Gloss* | *Morphosyntactic feature* |
|---|---|---|---|
| lauf-en | läuf-t | 'run' | 3SG present verb |
| Huhn | Hühn-er | 'hen' | plural |
| Vater | Väter-chen | 'father' | diminutive |
| Europa | europä-isch | 'Europe' | adjective formation |
| hoch | höch-st | 'high' | superlative |

In the Arawakan language Terena, first person singular is realized by progressive nasalization (from left to right), with (simplifying somewhat) the spread stopped by obstruents, which become prenasalized (see Akinlabi 2011 and references cited there):

(2) | 3SG | 1SG | *Gloss* |
|---|---|---|
| arıne | ãɾĩnẽ | 'sickness' |
| emoʔu | ẽmõʔũ | 'boss' |
| owoku | õw̃õⁿgu | 'house' |
| ıwuʔıʃo | ĩw̃ũʔⁿʒo | 'to ride' |
| takı | ⁿdaki | 'arm' |
| paho | ⁿbaho | 'mouth' |

In the Ethiopian Semitic language Chaha, verbs suffixed with the third singular masculine object marker (3SG.MASC.OBJ) show labialization of the rightmost labializable consonant (Banksira 2000; Rose 2007). The -*n* morpheme is analyzed as a "case" affix that precedes 3SG.MASC.OBJ, so that the middle column is derived from √ROOT-"CASE"-3SG.MASC.OBJ (3SG.MASC.OBJ position is marked with △):

(3) | *Without* OBJ | 3SG. MASC. OBJ | *Gloss* |
|---|---|---|
| kətəfə | kətəfʷə-n-△ | 'chop' |
| nəkəsə | nəkʷəsə-n-△ | 'bite' |
| k'əsərə | k'ʷəsərə-n-△ | 'erect' |

My primary claim is that essential generalizations about the locality of alternations like those seen in (1)–(3) follow directly in a morpheme-based theory, but not in an affixless theory. An initial statement of the observation to be explained in this way is given in (4):

(4) Morphophonological Locus (ML)

A morphophonological rule triggered by morpheme X behaves as if the effects of the rule are local to the position of X.

The wording in (4) assumes that there are morphologically conditioned phonological rules—that is, that the identity of morphemes is available in the phonology, such that phonological processes may be triggered by certain morphemes, or apply to some morphemes and not to others. See section 9.3 for some further discussion of this point.

The importance of Morphophonological Locus has surfaced in the literature in some different forms. For example, Lieber (1987), who develops a theory in which the exponent of a morpheme may be (or include) an autosegment, emphasizes that the locality of mutation processes (among which she includes German umlaut) follows from the position of a morpheme in a complex word. Other observations along these lines can be found as well.[2] However, these observations have not, to my knowledge, been organized into a general theory that emphasizes the centrality of the morpheme for nonaffixal morphology.

9.2.1 Morphemes and Morphophonological Loci

Morphemes play a defining role in explaining Morphophonological Locus. A starting point in the theory of this effect is the idea that in a morpheme-based theory, "words" are realizations of morphemes combined into syntactic structures; I will assume that these are complex heads of the type schematized in (5):

(5) $[[[\sqrt{\text{ROOT}} \ W] \ X] \ Y]$

Morphemes arranged in a structure like (5) are linearly ordered in the PF component of the grammar; this gives them a linear position with respect to each other. With this in mind, my claim is that the generalizations stated as Morphological Locus (4) are accounted for by the Morphological Locus Theorem (6):

(6) *Morphological Locus Theorem*

A morphophonological change triggered by morpheme X is phonologically or morphologically local to X.

The rest of this section shows how (6) follows as a theorem in a theory with (i) morphemes, along with (ii) morphological and phonological locality conditions on operations.

With respect to (ii), some care must be taken to explain why (6) makes reference to both *phonological* and *morphological* locality; this is a key theme throughout this section. For the moment, it can be seen how (6) accounts for the basics of (1)–(3). In the case of German umlaut, the affixes triggering the fronting process are suffixes. Thus the fact that the process applies to the final vowel of *Europa* to derive the adjective *europä-isch*—that is, that umlaut applies locally from right to left—follows directly.[3] With Terena 1SG, the change is triggered by (or is the exponent of) a prefixal agreement (AGR) morpheme: [AGR [Noun/Verb. . .]] (see, e.g., Akinlabi 1996, 2011; Wolf 2006); this explains why nasalization has the locus that it does. Although there is no "overt" prefixal material before the verb with 1SG, there is an overt 1PL AGR prefix with vowel-initial words, and 2SG AGR is marked morphophonologically from left to right as well (Ekdahl and Grimes 1964); there is thus clear motivation for prefixal AGR. Finally, the Chaha 3SG. MASC.OBJ morpheme originates in a suffixal position, where other object morphemes occur; the fact that it labializes locally from right to left is therefore expected.

The MLT (6) is at the core of a morpheme-based theory's morphophonological predictions, and, as simple as it is, it cannot be formulated straightforwardly in an affixless approach (section 9.3). Of course, some different auxiliary theories must be combined with the MLT to account for certain phenomena noted in the literature. For example, the infixation of a morpheme will result in that morpheme not being in its locus as defined by the MLT. However, as noted by Halle (1990), infixes are prefixes or suffixes that are subsequently moved, in my view either morphologically (see Embick and Noyer 2001, Embick 2007b) or phonologically (see, e.g., Halle 2001 and the overview in Yu 2004). The position of infixed morphemes is therefore expected to be local to their original position, as defined by the MLT. Effects that are perhaps similar because of their relation to phonologically defined objects are found with reduplication and other prosodic phenomena.[4] Although I cannot examine these phenomena here, they highlight the importance of taking the MLT as part of a theory that incorporates both morphological and phonological representations and locality conditions.[5]

9.2.2 Locality

The MLT specifies that morphophonological changes have a locus. Beyond this, there must be an additional theory of the precise locality conditions under which such alternations occur. Building on earlier work (see below), I hypothesize that there are two kinds of "morphophonological" alternation in the broad sense. One is subject to *morphological* locality, which I take to be the concatenation (= immediate linear adjacency) of morphemes; the other type obeys *phonological* locality (e.g., adjacency in autosegmental representations).

As a first step, some terminology is in order. Morphophonological alternations have *triggers* (the cause of the alternation) and *targets* (the object that undergoes the phonological change). Moreover, both targets and triggers can be either *morphologically* (M-) or *phonologically* (P-) defined. M-triggers are seen in (1)–(3), where specific morphemes induce the change; M-targets are found when a particular set of Roots or morphemes undergo a change (while others do not) and the set cannot be defined phonologically. For example, German umlaut applies to e.g. *laufen* 'to run' in the 3SG present tense to produce *läuf-t*; but e.g. *kaufen* 'to buy' does not undergo umlaut (3SG present *kauf-t*). Or, for instance, Spanish diphthongization occurs under stress, yielding alternations like *pensár* 'to think', *piénso* 'think-1SG'; but it only applies to certain targets like $\sqrt{\text{PENS}}$ and not others (compare *tensar* 'tauten', with 1SG *tenso*).[6] P-targets and triggers are defined in purely phonological terms—that is, without reference to the specific morphemes involved.

In these terms, Embick (2012) hypothesizes that there are two distinct types of morphophonological alternations:

(7) a. *Morpheme/Morpheme (M/M) Rules):* Rules that have an M-Target and an M-Trigger
        ⇒Operate in terms of morphological locality/Cannot skip morphemes.
     b. *Morphophonological Rules (M/P) Rules):* Rules that have either an M-Target or an M-Trigger, with the other component being phonological
        ⇒Operate in terms of phonological locality/Can skip morphemes.

Starting with M/M-Rules, Embick (2010) builds on earlier work in proposing that contextual allomorphy—crucially, the suppletive type effected by the Vocabulary Insertion operation—requires the concatenation of morphemes. More precisely, a morpheme X can have its allomorphy determined by Y only when it is immediately adjacent to Y—that is, when $Y\frown X$ or $X\frown Y$. Building on this proposal, Embick (2012) observes that the M/M-Rules (7a) require information about two morphemes in exactly the same way. For example, in the English pasttense, the Root $\sqrt{\text{SING}}$ appears as *sang* when it is local to the past-tense morpheme T[+past]. For this to happen, both $\sqrt{\text{SING}}$ and T[+past] have to be visible *as morphemes*—that is, as the specific Root and morpheme that they are—in order for the change to apply.[7] In a way that covers both M/M-Rules and contextual allomorphy, the *Morpheme Interaction Conjecture* hypothesizes that all processes referring to two morphemes as morphemes are subject to the same linear locality condition:

(8) *Morpheme Interaction Conjecture (MIC)*

> PF Interactions in which two morphemes are referred to *as morphemes* occur only under linear adjacency (concatenation).

Calabrese (2012) provides a compelling analysis of a collection of effects in the Italian past tense called the *passato remoto* that illustrates the effects of concatenation in M/M-Rules. This tense shows a number of irregular verbs with stem allomorphy restricted to the 1SG, 3SG, and 3PL forms; (9) illustrates with a small sample of such verbs:

(9)	INF	1SG	2SG	3SG	1PL	2PL	3PL
a. 'come'	venire	vénni	venísti	vénne	venímmo	veníste	vénnero
b. 'move'	mwovere	mɔ́ssi	mwovésti	mɔ́sse	mwovémmo	mwovéste	mɔ́ssero
c. 'put'	mettere	mísi	mettésti	míse	mettémmo	metttéste	mísero
d. 'see'	vedere	vídi	vedésti	víde	vedémmo	vedéste	vídero

The 1SG, 3SG, and 3PL forms of these verbs show changes to the verb stem, whereas the rest of the person/number combinations show the stem form found in other tenses (cf. the infinitives *venire*, etc.). The relevant changes are triggered by the past-tense morpheme T[+past], and apply only to certain verbs, making them M/M-Rules in the classification above. Calabrese shows that traditional explanations, which rely on (essentially suppletive) allomorphy driven by stress, fall short in explaining the distribution of regular and irregular stem alternants. His argument is that the irregular stem alternants are found only in the 1SG, 3SG, and 3PL because it is precisely these forms that have no theme vowel. According to this analysis, then, there are two morphological representations underlying the two type of Passato Remoto, athematic (10a) and thematic (10b):

(10) a. Athematic $\sqrt{\text{ROOT}}$ -T[+past]-AGR: 1SG mwov-s-i → mɔ́ssi
     b. Thematic $\sqrt{\text{ROOT}}$ -THEME-T[+past]-AGR: 2SG mwov-e-Ø-sti → mwovésti

The representations in (10) abstract away from certain details that play an important role in Calabrese's treatment (e.g., Tense and AGR fuse in thematic forms). The key point for present purposes is that the rules that derive irregular stem allomorphy are constrained to apply only when the Root and T[+past] are adjacent. As Calabrese demonstrates, this locality-based view accounts for stem allomorphy and a number of other morphophonological effects seen in the Passato Remoto forms in a direct and constrained way—a significant advance over alternatives in which stem choice is determined by paradigmatic structure or global phonological properties.

Moving past M/M-Rules and the MIC, for the M/P-Rules (7b) what is at issue is the idea that locality defined in terms of phonological representations may be morphologically "nonlocal" (i.e., may skip morphemes).[8] Thus, even though such rules are triggered by specific morphemes, or target specific morphemes, they obey the locality conditions that apply to phonological rules. This is an important observation, because it highlights the fact that such rules are "part of the phonology," despite their morphological conditioning.

A first illustration of the "morpheme-skipping" effect can be drawn from Banksira 2000, using the process of labialization in Chaha (recall (3) above). This process labializes the first potential target to the left of the 3SG.MASC.OBJ morpheme, whose position is symbolized with -ə△ below.[9] In the following examples, labialization affects the malefactive morpheme -β in (11a), skips benefactive -r to labialize 1SG subject -xɨ in (11b), and skips benefactive -r and 3SG.MASC.SUBJ -Ø in (11c) to labialize the medial consonant of the root. Note that all of these examples include a final tense morpheme -m that, even though it is phonologically a possible target of labialization, is to the right of the 3SG.MASC.OBJ morpheme, and therefore never targeted; this is a good example of Morphophonological Locus, since right-to-left labialization starts from the object morpheme, and not, for instance, from the right edge of the word (labialized elements are boldfaced):

(11) *From Banksira 2000, 284, 296–297*

    a. kəfət -xɨ-   -β   -ə△   -m → kəfət-xɨ-**w**-ə-m
        open -1SG.SUB -MAL -3SG.MASC.OBJ -TNS
        'I have opened (something) to his detriment.'

    b. kəfət -xɨ   -r   -ə△   -m → kəfət- **xʷɨ**-r-ə-m
        open -1SG.SUBJ -BEN -3.SG.MASC.OBJ -TNS
        'I have opened for him.'

    c. kəfət -Ø   -r   -ə△   -m → **kəfʷətə**-Ø-r-ə-m
        open -3SG.MASC.SUBJ -BEN -3SG.MASC.OBJ -TNS
        'He has opened (something) for him.'

A second example of skipping morphemes can be seen in some dialects of Italian, which show a process called *metaphony*: the raising of a stressed vowel when the following syllable contains a high vowel. For an overview see Maiden 1991 and, for the phonological change(s) effected, Calabrese 1999, 2009. In certain dialects of Italian, only schwa appears posttonically. In some of these dialects, the second-person singular agreement (2sG AGR) affix—which historically was (metaphony-triggering) -*i*, as in Standard Italian—continues to trigger metaphony. This is shown in (12) for the dialect of Ischia (the left columns show Standard Italian for comparison):

(12) *Metaphony triggered by AGR (Maiden 1991, 159)*; cant/kand '*sing*'

	St. Italian		Ischia, Campania	
	pr. ind.	*impf. ind.*	*pr. ind.*	*impf. ind.*
1sG	canto	cantavo	kandə	kandavə
2sG	canti	cantavi	**kɛndə**	**kandɛvə**
3sG	canta	cantava	kandə	kandavə

Two observations can be made here. The first is that the raising is a morphophonological change triggered by the 2sG AGR morpheme. The second is that the target of the change need not be morphologically adjacent. The imperfect indicative form *kandɛvə* consists of four morphemes: a Root, a theme vowel (underlyingly /a/), a past-tense morpheme -*v*, and the 2sG AGR morpheme -ə: [[[√KAND a] v] ə]. The change that is effected by metaphony triggered by 2sG AGR, then, is not restricted to adjacent morphemes. Rather, it skips the past-tense morpheme -*v*—that is, it obeys *phonological* locality and affects the autosegmentally adjacent theme vowel.

In both the Chaha and the Ischia examples, the morphophonological change may be realized on a morpheme that is not morphologically concatenated with the trigger of the change. Rather, it is realized on an element that is phonologically local to the trigger.

9.2.3 Synopsis

The theory outlined in this section hypothesizes that morphophonological changes in the broad sense might be the result of rules that have distinct locality conditions: M/M-Rules, which apply to concatenated morphemes, and M/P-Rules, which operate in terms of phonological locality.[10]

The full range of predictions of this approach remain to be investigated. Moreover, there is more work to be done on the precise nature of integrated morphological and phonological representations, in which specific morphemes and their linear relations play a role (along the lines investigated in, e.g., McCarthy 1981 and subsequent work). It should be stressed, however, that while many alternative formulations of morphophonological locality may be considered, the core fact to be accounted for is that there are *some* locality conditions that regulate the application of morphophonological alternations. That is, Morphophonological Locus must be accounted for, and theories are deficient to the extent that they allow stem-changing and other morphophonological alternations to be triggered in an unrestricted way. This is the essential theme of the next section, where comparisons with nonaffixal theories are undertaken.

9.3 Some Pertinent Comparisons

At the outset I noted that nonaffixal changes are often taken to be problematic for morpheme-based theories, and as evidence for affixless theories. The reasons for this are supposed to be clear—for example, whereas *-ed* in the past tense *played* of *play* looks like a morpheme, the change seen in *sang* from *sing* does not; not in any obvious sense, anyway. By this last comment I mean that it is probably true that if we look only at the form *sang* (or at the forms *sing* and *sang* together), it might not be obvious why *sang* should be analyzed as containing (at least) the two morphemes $\sqrt{\text{SING}}$ and T[+past]. On the other hand, if we consider the syntacticosemantic fact that *sang* (like e.g. *play-ed*) is used for "past tense of $\sqrt{\text{SING}}$" only in a subset of past-tense clauses in English—that is, if we consider its relation to clausal syntax and interactions with T-to-C movement, negation, *do*-support, and so on—the need to treat *sang* as consisting of $\sqrt{\text{SING}}$ and T[+past] at some level of analysis is much clearer. On this point, see Chomsky 1957, 58, which is framed with reference to Hockett's (1954) discussion of how morpheme-based theories might handle nonaffixal alternations.

In any case, although the idea that morpheme-based theories have difficulties with nonaffixal alternations is familiar, explicit arguments that elaborate and develop this point are not always easy to find. Anderson 1992 provides a useful point of reference, because it is a sustained attempt to motivate and develop a theory that dispenses with morphemes (for inflectional morphology, anyway). As a justification for this move, Anderson argues that "the class of contributors to the form of complex words looks more like the set of changes made by phonological rules than it does like a lexicon of listed word-like elements" (p. 72). Generalizing, the conclusion is that "since a process-based approach naturally accommodates affixation, but not vice versa, the alternative we should prefer is to explore a theory of morphological processes" (p. 68).

It is important to note that this argument concentrates on the question of whether nonaffixal alternations can be reduced *in toto* to affixation. This is, in my view, somewhat misleading. What is at issue is this. In a theory with morphemes, an apparently nonaffixal alternation like German umlaut could, in principle, be treated in one of two ways: either (i) (morpho)phonologically, with a list of morphemes that trigger a fronting rule; or, (ii) via insertion of autosegments, so that, for instance, umlaut-triggering morphemes are those whose exponents begin with the feature [−back] (see, e.g., Lieber 1987). In the latter case, the need for "morphologically triggered phonological rules" is avoided; instead, morphemes plus "normal phonology" produce what look like nonaffixal changes. For convenience, the generalized version of the type (ii) approach—one that replaces morphologically conditioned phonological rules with insertion— will be referred to as a *Vocabulary Insertion Only* (VIO) theory. In these terms, Anderson (1992, 68) criticizes Lieber (1987) (and others) on the grounds that while certain nonaffixal alternations might lend themselves to a VIO analysis, others (subtractions, exchange rules, chain shifts) cannot be treated in affixal terms. For this reason, Anderson concludes, an affixless theory is required.

It is important to note that the correctness of the VIO approach is irrelevant as far as the main argument of this chapter is concerned. The reason for this is that the argument centered on Morphophonological Locus can be framed either in terms of a theory with morphologically triggered phonological rules (as in section 9.2), or in terms of a theory with VIO (since the inserted autosegments, etc., will have their locus in a morpheme that has a position). For the rest of this section, then, the question to be addressed is not *Can all "morphological changes" in the broad sense be treated affixally?* Instead, it is *Do morpheme-based or affixless theories provide the basis for the correct theory of morphological and morpho-phonological locality?* The status of VIO will be left for another occasion.[11]

I will speak generically of affixless theories in addressing the locality question, so that the argument is intended to apply to a wide range of approaches.[12] Schematically, affixless approaches derive the forms of words by operations that apply to representations consisting of Roots and matrices of features like (14). Compare the structured complex of morphemes in (13):

(13) $[[[\sqrt{\text{ROOT}} \pm W] \pm X] \pm Y]$

(14) $\sqrt{\text{ROOT}} \begin{bmatrix} \pm W \\ \pm X \\ \pm Y \end{bmatrix}$

In the affixless representation in (14), all of the features are equally "close" to each other and to the Root. This is a defining property of such a theory. With this in mind, the claim I will now develop is stated in (15):

(15) Affixless theories make no predictions about the locality of morphophonological interactions, because the MLT (or something similar) cannot be formulated in such theories.

Comparison with the theory of section 9.2 is important for understanding (15). Take, for concreteness, an instantiation of (13)/(14) in which *W* is Aspect, *X* is Tense, and *Y* is Agreement; this is the typical structure of a verb in many languages. Assume further that these are suffixes: $\sqrt{\text{ROOT}}$-ASP-TNS-AGR. The theory of section 9.2 says that in complex tenses, in which there are overt realizations Aspect and Tense, a morphophonological rule triggered for example by 1PL AGR might not be able to affect the Root. If it is an M/M-Rule, then it is predicted that no change to the Root will be possible, period, because such rules require the concatenation of morphemes. If the change is effected by an M/P rule, then it could only affect the Root if the exponents of Aspect and Tense are not themselves phonological targets (or phonological blockers) of the change.[13]

The point of (15) is that these types of predictions cannot be made in a theory that eliminates morphemes. There are at least two ways of making this position precise, corresponding to two different ways of analyzing nonaffixal changes in affixless theories.

A first type of analysis employs *stem insertion*. Anderson's (1992) treatment of irregular inflection and stem allomorphy holds that, for instance, *sang* is a (suppletive) allomorph ("stored stem") of $\sqrt{\text{SING}}$; the [+past]-specified stem *sang* beats the default stem *sing* when $\sqrt{\text{SING}}$ is bundled with T[+past].[14] Stem insertion has serious problems with blocking, as discussed by Halle and Marantz (1993). The further point raised by (15) is that it makes it impossible to formulate a local theory of Morphophonological Loci: since all of the features are equally local in representations like (14), any feature (or any subset of features in a representation like (14)) could potentially trigger stem allomorphy of a "nonaffixal" morphological type. The idea that certain changes cannot occur because of the distance between the trigger and target cannot be formulated. Thus, in this approach, there are not expected to be any effects of morphophonological locality in language.[15]

A second way of reasoning through (15) is as follows. Theories like Anderson's (e.g., Stump 2001, although it differs in other ways) use blocks of rules to spell out feature bundles like the one in (14). In Anderson's formalization, these *word formation rules* (WFRs) are phonological rewrite rules. So, for example, the plural of an English noun like *dog* is formed by taking the representation *dog*[+pl] and rewriting it with the rule /X/+pl →/Xz/ that adds /z/. Suppose now that rather than treating nonaffixal changes with stem insertion, as discussed above, such changes were treated with rewriting rules. Again, the question is as follows: if all features are equally close to the Root and to each other in representations like (14), then why should morphophonological changes ever show any sort of locus? The representation in (14) makes any potential interaction among features possible, thus allowing every conceivable trigger/target interaction. So, "local changes" could be analyzed by manipulating how the rule blocks are ordered. Letting B_W, B_X, and B_Y be rule blocks realizing W, X, and Y, it would be possible to say that the feature Y triggers a change that acts upon the output of B_W and B_X applying to the Root by stipulating the block order (i) B_W, (ii) B_X, and (iii) B_Y. But this order in no way follows from (14), where, as stressed above, all features are equally close to each other. Thus, it would also be possible to order a block $B_y{'}$ sensitive to Y first (since Y and the Root are visible to each other in (14)), such that changes triggered by the Y feature would apply to the Root even when there are overt reflexes of W, X, and Y in that order. More generally, the morphosyntactic representation (14) does not place any constraints on the order of morphophonological rule blocks, making any set of interactions possible. In short, affixless theories may manipulate rule blocks to account for local alternations, but only because they allow for completely nonlocal alternations as well.

Another way of making this point is as follows. In an affixless approach that employs rewrite rules, such rules are not expected to have the locality properties of phonological rules, because they are morphological—that is, triggered by features in representations like (14). But since there are no morphemes, they do not have morphological locality properties either. Anderson (1992, 45–46) seems to be aware of this issue, but does not, in my view, sufficiently acknowledge its implications, namely, the predicted absence of any locality effects in morphophonology.

In summary, to the extent that extremely nonlocal effects of the type outlined above are not found, nonaffixal changes are problematic for affixless theories. Given the observations about Morphophonological Locus outlined above, the burden of proof must be on advocates of the affixless theories to show either that (i) there are in fact radically nonlocal morphophonological changes in the world's languages, or that (ii) there is a straightforward way of accounting for morphophonological locality in an affixless theory.

9.4   Conclusions and Further Directions

The basic claims of this chapter are that (apparently) nonaffixal morphological changes (i) have a Morphophonological Locus in a word that determines where they apply, and that (ii) the correct theory of such Morphophonological Loci follows from a morpheme-based theory, but cannot be derived in an affixless view. Within the general framework of assumptions that I have adopted here, different approaches can be taken toward the locality of morphophonological operations. At a minimum (and putting aside the question of reducing every-thing to Vocabulary Insertion), there are M/P-Rules that show the locality conditions characteristic of phonological operations. Such rules connect with important developments in generative phonology, in which morphologically conditioned phonological rules are treated with other "normal" phonological rules (Halle 1959 and related work). For reasons that connect with the type of information found in (suppletive) contextual allomorphy, I hypothesized further in section 9.2 that there might be another type of rule, the M/M-Rules, which, because they make reference to the identity of two morphemes as morphemes, require morphological concatenation.

There are many additional topics to be addressed in a more comprehensive theory of morphophonology. Fundamental representational questions about how morphological and phonological information is accessed in the PF com-ponent connect with other substantive questions about the division of labor between Vocabulary Insertion and the phonology (leading, for example, to the question of how much "nonaffixal" morphology can be reduced to the former). Whatever specific directions these latter lines of investigation take, the general point that defines the present work is that the morpheme is indispensible for understanding how syntax, sound, and meaning are connected in language. In this I echo Halle 1990, a paper that has launched so much productive work because of its insistence that the morpheme in all of its dimensions must be at the heart of morphological theory.

Notes

Some of this material was discussed in my spring 2012 seminar at Penn, and I thank the participants for a number of important comments, suggestions, and corrections. Thanks also to two reviewers for a number of detailed and helpful comments that have greatly improved the chapter; I regret not having the space to address their points in full. The author's research is supported by the Eunice Kennedy Shriver National Institute of Child Health & Human Development of the National Institutes of Health under Award Number R01HD073258.

1. I say "evidently" here because if autosegments can be the exponents of Vocabulary items, then at least some of these alternations could be treated with "normal" Vocabu-lary Insertion (see section 9.3).

2. For example, Salanova's (2004) study of truncation points to the role of morphemes in constraining morphophonological changes. Elsewhere in the literature, the idea that various morphophonological changes are effected by autosegments etc. that must be either prefixal or suffixal can be found in Akinlabi 1996 (Akinlabi (2011) calls this property "directionality"). Along the lines pursued by Lieber, Wolf (2006) employs constraints that force mutations to have a locus and argues against affixless versions of OT morphophonology on this basis.

3. See also Lieber 1987 and Wiese 1996a for this point. For some discussion of the phonological locality of this process in forms like *Väter-chen* see, for example, Kiparsky 1996.

4. With reduplication, this is sometimes analyzed with the idea that "heads" are targeted, as in Aronoff 1988. Other phenomena worth examining in this connection involve "augmentation" of the types seen in Classical and Modern Greek; it is also possible that *ge-* prefixation in German participles could be analyzed in these terms.

5. In fact, morphological and phonological locality are only part of the picture. For reasons discussed in Embick 2010 with reference to allomorphy, syntactic locality (the theory of phases, Chomsky 2000, 2001; see Marantz 2000, 2007, and chapter 6, this volume, as well as Embick and Marantz 2008) also plays a role in certain morphophonological interactions. However, integrating such considerations into morphophonological theory presents numerous complications, as stressed by Lowenstamm (2010) with reference to the Level 1/Level 2 distinction in English affixes. See as well Marvin 2002 and chapter 5, this volume.

6. The relationship between affixation, stress, and diphthongization in Spanish is quite complicated. For example, there are well-known cases in which certain affixes that affect stress do not affect diphthongization; see Bermúdez-Otero 2006 and references cited there, as well as Embick 2012 for some comments from the perspective of the framework discussed here.

7. Note in addition that for this to occur, the T[+past] morpheme and $\sqrt{\text{SING}}$ also have to be in the same phase-cyclic domain; see Marvin 2002 and Embick 2010 for discussion.

8. Carstairs-McCarthy (1992) highlights the importance of "morphologically nonlocal" interactions along these lines, with an illustration from Zulu palatalization. See also Hyman, Inkelas, and Sibanda 2008 for some related phenomena.

9. Banksira treats 3SG.MASC.OBJ as -∂U, where the /U/ component contributes [round] and [high] features that are spread to the left. See his book for other important details concerning the phonology of labialization.

10. It is also conceivable that certain alternations might actually be triggered in both ways. With respect to German umlaut, a reviewer makes the important observation that while certain affixes trigger the change in a target-specific way (recall the examples in section 9.2.2), other affixes appear to trigger it regularly. These are called "umlaut variable" and "umlaut conditioning" respectively in Lieber 1987, 100. One possible line to investigate is that umlaut is an M/M-Rule with the former class of affixes, but an M/P-Rule with the latter. As the reviewer notes, further complications arise because of apparent cases of optionality in the umlaut system. In any case, much remains to be said about this process with reference to the M/M- versus M/P-Rule classification, but considerations of space rule out further discussion here.

11. Regarding VIO for M-Triggers, Bye and Svenonius (2012) develop something along the lines of Lieber's (1987) program (although other assumptions that they make about insertion at nonterminals might complicate the predictions about Morphological Locus; see note 15 below). It is not clear at this point that VIO extends naturally to all of the phenomena treated with morphologically conditioned phonological processes. Beyond the question of how all M-Triggers can be reduced to Vocabulary

Insertion, a further question for VIO is how to account for the properties of M-*Targets*. Recall from section 9.2.2. that processes like German umlaut and Spanish diphthongization apply to some morphemes and not to others. Illustrating with the latter, the morphemes undergoing this process could be identified morphologically (e.g., with diacritics, as in Harris 1969), or phonologically (by making the underlying phonological representations of diphthongizing $\sqrt{\text{PENS}}$ and nondiphthongizing $\sqrt{\text{TENS}}$ distinct, as in Harris 1985). To the extent that "abstract" phonological solutions are not always available (or correct) for M-Targets, the theory will require phonological rules that make reference to specific morphemes.

12. For example, there are several affixless approaches that, like Anderson, follow the lead of Matthews 1972, Aronoff 1976, and others (e.g., Pullum and Zwicky 1991 and Stump 2001. With respect to stem alternations in particular, there is also the "morphomic" approach advocated in Aronoff 1994 (also a continuation of views from Matthews 1965, 1972), which has connections as well to diachrony (e.g., Maiden 2004a; see many of the papers in Maiden, Smith, Goldbach, and Hinzelin 2011). Many of these movements separate themselves from broader architectural questions concerning syntax, semantics, phonology, and so on in such a way as to make substantial comparisons difficult (although see Embick 1998, 2000, and Embick and Halle 2005 for some discussion of Aronoff's approach).

On the more experimental side, work in the "words-and-rules" and related frameworks seems to assume something like Anderson's view of what it means to be formed by rule; see, for instance, Pinker 1999, Pinker and Ullman 2002, and the discussion in Embick and Marantz 2005. Further afield in terms of theoretical perspective, Seidenberg and Gonnerman 2000 and Hay and Baayen 2005 are representative examples of approaches that seek to eliminate morphemes in more radical ways.

13. If the M/P rule is iterative, it could apply to intervening Aspect and Tense, and the Root as well.

14. The same kind of stem insertion could be at the heart of Maiden 2004 and related treatments of stems, although it is difficult to tell, since an insertion mechanism is not specified.

15. This argument also applies to approaches like Siddiqi 2009, which treats stem allomorphy of the *sing/sang* type by *fusing* nodes in structures like $[[\sqrt{\text{SING}}\,v]\ \text{T[+past]}]$; this creates representations like (14) prior to Vocabulary Insertion. Some theories that allow insertion of phonological material at nonterminal nodes are subject to this argument as well (see Bye and Svenonius, 2012 for references), to the extent that the relevant nonterminals contain feature bundles like (15).

10 Agreement in Two Steps (at Least)

Eulàlia Bonet

10.1 Agreement Asymmetries

In many languages the elements of a DP agree with a head noun (nominal concord). This is illustrated in (1) with an example from Spanish, where there is gender and number concord with the head noun, which appears in boldface.

(1) Est<u>as</u> pequeñ<u>as</u> **cas<u>as</u>**   abandonad<u>as</u>
  this.FEM.PL small.FEM.PL house.**FEM.PL** abandoned.FEM.PL
  'these small abandoned houses'

However, concord seems to fail in many languages for some positions, a phenomenon that has been called *lazy concord* (see Haiman and Benincà 1992; Rasom 2008). In most cases, lazy concord affects prenominal elements, not postnominal ones, as the example in (2) from Moroccan Arabic illustrates (data from Shlonsky 2004). As (2a) shows, demonstratives in postnominal position agree in gender and number with the noun, in boldface; in (2b) the demonstratives in prenominal positions are bare forms.

(2) a. l **wəld** had-a   l **bənt** had-i   lə **wlad** had-u
   the boy this.MASC.SG the girl this.FEM.SG the children this.MASC.PL
   'this boy'   'this girl'   'these boys'
  b. had l **wəld**   had l **bənt**   had lə **wlad**
   this the boy   this the girl   this the boys
   'this boy'   'this girl'   'these boys'

Some other languages that exhibit prenominal-postnominal asymmetries are Abkhaz (Hewitt 1979), with number concord postnominally but a bare form prenominally; Central Ladin (Rasom 2008), with feminine plural concord postnominally but only feminine concord prenominally; or Asturian

(Fernández-Ordóñez 2007), with mass concord postnominally but gender concord prenominally.

Agreement asymmetries have also been observed at the clause level (clausal agreement). In these cases the order subject-verb often causes full agreement while with postverbal subjects lazy agreement is found. An example appears in (3), from Standard Arabic (Aoun, Benmamoun, and Sportiche 1994). In (3a) the order S–V causes gender and number agreement on the verb, while in (3b) the order V–S causes only gender agreement on the verb; the subject appears in boldface.

(3) a. **ʔal-ʔawlaad-u**   naam-uu
      the-children.NOM   slept.3.MASC.PL
      'the children slept'

   b. Naam-a   **l-ʔawlaad-u**   (*naam-uu **l-ʔawlaad-u**)
      slept.3.MASC.SG   the-children.NOM
      'the children slept'

More examples and a typology of agreement asymmetries with postverbal subjects can be found in Samek-Lodovici 2002. Similar asymmetries have been observed in language acquisition (see, for instance, Franck, Lassi, Frauenfelder, and Rizzi 2006).[1]

Agreement asymmetries at the clause level and at the DP level show a clear parallelism: more stable agreement is found when the trigger of agreement precedes the element(s) it agrees with; less stable agreement is found when the trigger follows the target(s). This is schematically shown in (4), where the trigger appears in boldface. Throughout this chapter, the term *modifier* is used informally to refer to any element that can enter into an agreement relation with the noun within the DP.[2]

(4) a. *More stable agreement*   Clause level: **subject**   verb
                            DP level: **noun**   modifier
   b. *Less stable agreement*   Clause level: verb   **subject**
                            DP level: modifier   **noun**

10.2 Previous Analyses of the Asymmetries

Several accounts exist of agreement asymmetries within the clause or within the DP, and in some cases the suggestion is made that the mechanism that causes DP and clausal asymmetries is the same. Franck et al. (2006), following work in Guasti and Rizzi 2002, suggest that, at the clause level, full agreement between a preverbal subject and the verb is the result of features being checked twice, through Agree and Move (which gives rise to a spec-head configuration); the weakness found with postverbal subjects is due to features being checked only once, through Agree. Shlonsky (2004) and Nevins (2011a) claim that these two different mechanisms for agreement are the source of the asymmetries within the DP, the harder issue being how to relate postnominal concord to a spec-head relation. For Shlonsky (2004) full concord within the DP is achieved indirectly. The head X in (5), which has the NP as its complement, has semantic features associated with its specifier (AP), but it also has phi-features that trigger the movement of the NP to the specifier position of the AgrXP projected by the raised head X (see Shlonsky 2004 for details); spec-head agreement, direct or indirect, never arises between an Agr-bearing head and prenominal modifiers.

(5) $[_{\text{AgrXP}}$ NP $[$ AgrX^0 $[_{\text{XP}}$ AP $[$ t_X t_{NP} $]]]]$

Nevins (2011a) argues for a "roll-up" analysis of DP-internal movement (Cinque 2005), with evidence from DPs with two adjectives, and provides an analysis of concord based on simplified structure that omits possible DP-internal functional projections. As shown in the derived structure in (6), an NP that has raised to the specifier position of the most embedded adjective A2 agrees with it through a spec-head configuration: When this whole structure A2P is moved to the specifier position of a higher adjective A1, agreement is obtained again between the specifier A2P and the head A1. Prenominal modifiers are out of the scope of this roll-up movement; no agreement through spec-head is then possible.[3]

(6) $(_{\text{A1P}} (_{\text{A2P}} \text{NP} (_{\text{A2'}} \text{A2} \, t_{\text{NP}})) (_{\text{A1'}} \text{A1} \, t_{\text{A2P}}))$

In what follows I will concentrate on two very different proposals that, contrary to the ones summarized so far, not only suggest a reason for the existence and the direction of the asymmetries but also try to explain the crosslinguistic variation that is found, both with respect to the presence or absence of asymmetries in a given language and with respect to the particular features that might be involved in the phenomenon. These two proposals are Samek-Lodovici 2002 and Ackema and Neeleman (A&N) 2003. Both of them focus on clausal asymmetries, but can be extended to DP asymmetries.

Samek-Lodovici (2002) proposes to account for the asymmetries within Optimality Theory resorting to two universal constraints related to agreement. One of them, AGR$_f$, favors local agreement, spec-head agreement for the cases he considers. The other agreement-related constraint, EXTAGR$_f$, favors general agreement, both local and long distance. These two constraints compete with the constraint NoFEATS. The definition of these three constraints, all of them markedness constraints, appears reproduced in (7) (Samek-Lodovici 2002,

(19) (21), (22)). The constraint NoFeats, in spite of its name and its definition, has to be understood as a constraint that bans agreement. According to Samek-Lodovici it is violated when agreement occurs.

(7) a. NoFeats: No agreement features.
   b. AGR_f: An agreement head H and a DP must agree on feature f within the local projection HP.
   c. ExtAgr$_f$: An agreement head H and a DP must agree on feature f within the extended projection of H.

The ranking ExtAgr$_f$ >> NoFeats causes full agreement. The ranking NoFeats >> AGR_f, ExtAgr$_f$, has as a result a generalized lack of agreement. Agreement asymmetries are obtained when the spec-head agreement constraint AGR_f dominates the constraint against the expression of agreement features, NoFeats, and this one in turn dominates the constraint favoring general agreement, ExtAgr$_f$. The constraints AGR_f and ExtAgr$_f$ can be relativized to specific agreement features (e.g., AGR_{gender}, AGR_{number}, AGR_{person}), so a much finer typology can be derived, as shown by Samek-Lodovici (2002, (2), (3)).

The redefinition of these constraints, which is needed to extend the proposal to DP asymmetries, depends on the precise DP-internal structure and operations assumed, and it will not be attempted here; it is not a trivial matter. But the basic idea would be the same: one constraint controls local agreement only (AGR_f), and another constraint controls general agreement, local and nonlocal, within the DP (ExtAgr$_f$). Following the insights in Shlonsky 2004 and Nevins 2011a, AGR_f would penalize the lack of agreement between the head N (or, rather, NP) and other elements with which it enters, directly or indirectly, a spec-head configuration—that is, postnominal modifiers. The constraint ExtAgr$_f$ would cover agreement either through spec-head or Agree.

The tableau in (8) shows with an abstract example the asymmetries that we saw illustrated in (2) with an example from Moroccan Arabic. The input contains a prenominal modifier X (a determiner, an adjective), the head N with a feature [PLURAL], and a postnominal modifier Y.[4] Four candidates are considered: the winning candidate (8a), with only postnominal concord, a candidate with no concord (8b), a candidate with full concord (8c), and a candidate with only prenominal concord. The ranking AGR_{num} >> NoFeats >> ExtAgr$_{num}$ forces postnominal concord. Under this model, prenominal-only concord cannot be obtained under any ranking because that candidate will always be harmonically bounded by the candidate with postnominal-only concord. Candidates with this type of asymmetric concord are ignored in other tableaux.

(8)

Input: X N Y [PL]	AGR_{num}	NoFeats	ExtAGR_{num}
a. ☞ X N Y [PL] [PL]		*	*
b. X N Y [PL]	*!		**
c. X N Y [PL] [PL] [PL]		**!	
d. X N Y [PL] [PL]	*!	*	*

In Samek-Lodovici's proposal, both agreement and the agreement asymmetries are obtained in a single step. In section 10.3 it will be shown that a single-step approach of this type is not adequate for dialects of Spanish where concord asymmetries affect only a restricted set of nouns.

A&N (2003) (and also A&N 2004, 2012) argue for a very different model to account for agreement asymmetries (agreement weakening, in their own terms) within the clause. In their proposal, there is full agreement in the overt syntax, the weakening arising at PF within prosodic phrases. At PF, before the insertion of phonological material, there is an initial prosodic phrasing that, following work by Selkirk and other authors, is determined by alignment conditions that insert a prosodic phrase boundary ϕ at the right (or left) edge of each XP.[5] For head-initial languages the relevant constraint is as formulated in (9) (see, for details, A&N 2012, (3); A&N 2003, (4); A&N 2004, 186 (4)).

(9) Align the right edge of an XP with the right edge of a ϕ.

In (10) (corresponding to A&N 2003, (40)) it is shown how the initial prosodic phrasing between subject and verb differs in V-S (10a, a′) and S-V (10b, b′) sequences. The syntactic structures in (10a,b) map onto the initial prosodic phrases in (10a′,b′), respectively. The boundaries of prosodic phrases are indicated by curly brackets.

(10) a. $[_{FP} [_F V] [_{IP}$ subject $t_V [_{VP} t_V$ object$]]]$
    a′. {V subject} {object}
    b. $[_{FP}$ subject $[_F V] [_{IP} t_{subject} t_V [_{VP} t_V$ object$]]]$
    b′. {subject} {V object}

With V-S order the subject and the verb end up in the same prosodic phrase (see (10a′)), while with S-V order the subject and the verb end up in different prosodic phrases (see (10b′)). To account for the fact that in Standard Arabic, as illustrated in (3), there is a loss of number agreement between the subject and the verb in V-S order (not in S-V order), A&N (2003) propose the weakening rule in (11) (see A&N 2003, (41)).

(11) *Arabic agreement weakening*
$\{[V \ \text{PL} \ . . .] \ [D \ \text{PL} \ . . .]\} \rightarrow \{[V \ . . .] \ [D \ \text{PL} \ . . .]\}$

This weakening rule is a rule of allomorphy that is subject to a recoverability condition: "rules of suppression operate under agreement" (A&N 2003, (13)). They can apply to V-S sequences because the verb and the subject agree and belong to the same prosodic domain. Therefore the plural feature of the first terminal can be deleted without plurality being lost in that domain. In S-V sequences the rule cannot apply because the prosodic domain of the subject does not contain another terminal, so the context of the rule is not met. A&N (2003) propose very similar rules for other agreement asymmetries within the clause. Phonological material (the Vocabulary in Distributed Morphology terms, Halle and Marantz 1993) is inserted after rules of allomorphy have applied, and prosodic phrases can be modified to satisfy weight considerations, for instance.

But if agreement-weakening phenomena within the clause are to be attributed to PF rules that operate within prosodic constituents in the context of another terminal, one would expect agreement-weakening phenomena within the DP to be accounted for in the same fashion. For a complex DP with a prenominal adjective, a postnominal one, plus a couple of other prenominal modifiers, the prosodic structure would have to be as shown in (12).

(12) {Dem Num A N} {A}

This prosodic structure can only be obtained if there is only one right edge of an XP to the right of N (and after the second A); there cannot be any other right edges of XPs before the noun, because they would trigger the insertion of a prosodic boundary. The required syntactic configuration is incompatible with the analysis of DPs proposed in Shlonsky 2004, Cinque 2005, 2010, and much other work, because it would imply, for instance, that the specifier positions preceding the NP cannot contain any XP. To push the argument further, I will assume that some syntactic framework can allow the generation of the initial prosodic structure in (12).

Given the initial prosodic structure in (12), the prediction is that allomorphy rules would be able to weaken Dem and Num in (12) but not a postnominal A. Such an approach could account for gender and number loss in the example from Moroccan Arabic in (2). A simplified weakening rule deleting the two features could have the form in (13).

(13) $\{[X \ \text{FEM PL} \ . . .] \ [N \ \text{FEM PL} \ . . .]\} \rightarrow \{[X \ . . .] \ [N \ \text{FEM PL} \ . . .]\}$

This rule would delete gender and number features in prenominal modifiers, because they would belong to the same prosodic phrase as N, which contains the same features, but it would not affect postonominal modifiers, because they would constitute separate prosodic phrases.

10.3   A Challenge for Samek-Lodovici 2002 and A&N 2003: Gender Concord in Dialects of Spanish

The Spanish feminine definite article *la* surfaces as *el* before certain feminine nouns that start with a stressed *a*. So, feminine nouns like *agua* [áɣwa] 'water' (cf. *agua fría* 'water.FEM cold.FEM') surface with the definite article *el*, *el agua* 'the water', in spite of the fact that the regular feminine definite article is *la* (cf. *la nieve* 'the snow', *la imagen* 'the image', *la almohada* [almoáða] 'the pillow'). The change in the article does not affect categories other than nouns (not adjectives or adverbs: *la antes mencionada* 'the.FEM before mentioned.FEM'; not proper names: *la Ágata* 'the Agatha'), it applies only under adjacency between the definite article and the noun (*la plácida agua* 'the.FEM quiet.FEM water.FEM'), and it applies only with singular, not plural, nouns (*las aguas* 'the.FEM.PL waters.FEM.PL'). Some studies, like Halle, Harris, and Vergnaud 1991, have focused on the fact that in some dialects the replacement also occurs when the initial *a* is not stressed on the surface, as in *el agüita* [aɣwíta] 'the water (DIM.)' or *el aguanieve* [aɣwanjéβe] 'the sleet'. Others have tried to establish whether the article *el* used with feminine nouns is a surface allomorph of the feminine article (with deletion of the final vowel and insertion of an initial epenthetic vowel) or the masculine definite article, and some have concentrated on the segments that trigger the change, a contact between two low vowels, which would occur if the feminine article surfaced (cf. **la agua*). For different views on these issues see, among many others, Zwicky 1985, Harris 1987, Kikuchi 2001, and Cutillas 2003.

In other varieties of Spanish the change in the definite article has extended to the indefinite article, some quantifiers, and demonstratives. But here I want to concentrate on some varieties of Spanish that have generalized the use of masculine agreement: all elements that precede feminine nouns like *agua* in the DP are masculine, while all the postnominal ones are feminine. Some examples appear in (14) (for more examples, see Eddington and Hualde 2008 as well as Bonet, Lloret, and Mascaró, forthcoming). All prenominal modifiers are masculine while postnominal ones are feminine. The noun appears in boldface.

(14) a. el   nuevo   **arma**   secreta
        the.MASC new.MASC weapon.FEM secret.FEM

   b. todo   su   **área**   delantera
        all.MASC POSS.   area.FEM   front.FEM

   c. el   mismo   **agua**   parecerá fría
        the.MASC same.MASC water.FEM   will-seem cold.FEM

The three examples in (14) show that the masculine surfaces even if no vowel contact would occur. In (14a), for instance, the feminine article *la* would be next to a consonant: **la nuevo*. (14b) shows, in addition, that adjacency between the prenominal element and the noun is not a requirement: *todo* surfaces as masculine even though the adjacency with the noun is broken by an invariable possessive *su*. A valid conclusion for this variety is that the class of feminine nouns that trigger the concord asymmetry is now an idiosyncratic class of nouns that only historically triggered a phonological dissimilation process. Outside of the DP, nouns like *agua* trigger feminine agreement in all varieties of Spanish (*el agua parece fría* 'the water seems cold.FEM'; *la bebo* '(I) drink it.FEM'). The fact that the agreement asymmetry arises only for a subset of feminine nouns indicates that it is not a purely syntactic phenomenon.

Assuming that masculine is a default gender, and ignoring for the time being the fact that the asymmetries in gender concord arise only in the singular (not in the plural), a general asymmetry between prenominal and postnominal elements could easily be accounted for in Samek-Lodovici's model, as illustrated in (15) with the example in (14a). This tableau is a concrete example of the schematic tableau in (8), and the same type of candidates are considered. Only the N *arma* is inherently specified as feminine in the input; the constraint hierarchy favors the candidate with postnominal, but not prenominal, concord, due to the higher ranking of AGR_{gen}, the constraint that forces local agreement, over ExtAGR_{gen}, the more general agreement constraint.[6]

(15) Tableau corresponding to *el nuevo arma secreta* 'the new secret weapon'

Input:	1	nuev-	arma [FEM]	secret-	AGR_{gen}	NoFeats	ExtAGR_{gen}
a. ☞	el	nuevo	arma [FEM]	secreta [FEM]		*	**
b.	el	nuevo	arma [FEM]	secreto	*!		***
c.	la [FEM]	nueva [FEM]	arma [FEM]	secreta [FEM]		***!	

One first question to address is how to restrict this asymmetry in concord to specific lexical items, because it is not the case that any feminine noun of the language can trigger the asymmetry. A paradox arises because for the quite restricted set of nouns that trigger this phenomenon we need the ranking shown in (15), $\text{AGR}_{gen} \gg \text{NoFeats} \gg \text{ExtAGR}_{gen}$, while for the vast majority of nouns, which show full concord, we need the opposite ranking, ExtAgr_{gen} >> NoFeats (the ranking of Agr_{gen} being irrelevant). In optimality theory two main proposals have been made to account for lexically restricted processes: cophonologies (proposed in Orgun 1996 and Anttila 1997) and lexically indexed constraints (developed by Itô and Mester 1999 and Pater 2000). Under the first of these two approaches, as instantiated in Inkelas and Zoll 2007, specific lexical items are subject to specific constraint rankings; they have different subgrammars. In the case at hand, the language would have a general ranking Agr_{gen} >> {ExtAgr_{gen}, NoFeats}. The set of items that include *agua* would be lexically specified for the ranking NoFeats >> ExtAgr_{gen}, while the complementary set of items would be specified for the opposite ranking: ExtAgr_{gen} >> NoFeats. Although this approach would work mechanically, it faces a problem pointed out in Pater 2010, namely that it cannot differentiate variation from exceptionality. Here it is clear that items like *agua* are exceptional; all other nouns are subject to the default ranking ExtAgr_{gen} >> NoFeats. This ranking should not have to be specified for each lexical item, including new coinages. Under the lexically indexed constraints approach, certain constraints have a general version, C1, and a version that is restricted to affect only a class of lexical items L indexed for that constraint, $C1_L$. The application of a process to a restricted set of items is obtained through the ranking $C1_L$ >> C2 >> C1. The facts from Spanish can be accounted for if the lexically indexed constraint is NoFeats. The ranking of all the constraints will then be Agr_{gen} >> NoFeats_L >> ExtAgr_{gen} >> NoFeats.[7] With this ranking, postnominal concord is systematic. NoFeats_L is relevant only for nouns like *agua* or *arma*, for which it will prevent any prenominal concord features from being realized, as shown in (16). For the rest of the nouns of the language, ExtAgr_{gen} will favor across-the-board concord, as shown in (17).

(16) Tableau corresponding to *el nuevo arma secreta* 'the new secret weapon', with lexically indexed constraints

Input: 1 nuev- arma$_L$ secret- [FEM]	Agr_{gen}	NoFeats_L	ExtAgr_{gen}	NoFeats
a. ☞ el nuevo arma$_L$ secreta [FEM] [FEM]		*	**	*
b. el nuevo arma$_L$ secreto [FEM]	*!		***	
c. la nueva arma$_L$ secreta [FEM] [FEM] [FEM] [FEM]		***!		***

(17) Tableau corresponding to *la nueva policía secreta* 'the new secret police'

Input: 1 nuev- policía secret- [FEM]	AGR_{gen}	NOFEATS_L	EXTAGR_{gen}	NOFEATS
a. el nuevo policía secreta [FEM] [FEM]			**!	*
b. el nuevo policía secreto [FEM]	*!		***	
c. ☞ la nueva policía secreta [FEM][FEM] [FEM] [FEM]				***

The second question to address is how the two approaches to exceptionality can account for the fact that, as mentioned earlier, the concord asymmetry in Spanish affects only singular nouns, not plural nouns, as illustrated in (18).[8]

(18) a. el   nuevo   arma   secreta
        the.MASC.SG   new.MASC.SG   weapon.FEM.SG   secret.FEM.SG
     b. las   nuevas   armas   secretas
        the.FEM.PL   new.FEM.PL   weapon.FEM.PL   secret.FEM.PL

One difficulty stems from the fact that both cophonologies and lexically indexed constraints always target specific lexical items, not lexical items under certain conditions (here being exceptional only when singular). But here for both models some reference to the context would have to be introduced. Within cophonologies nouns like *arma* would need to have two different constraint rankings contextually determined in their lexical entry. For instance, *arma* would have, in its lexical entry, the specifications EXTAGR_{gen} >> NOFEATS / __ PL and NOFEATS >> EXTAGR_{gen}/ __ SG; this specification would not capture the observation that, when plural, nouns like *arma* behave like any normal noun in the language. In the lexically indexed constraints approach the constraint NOFEATS_L would have to be restricted to apply when the noun is singular; the constraint would be something like NOFEATS_L / __ SG. A second difficulty comes from the vagueness of the definition of the constraint NOFEATS. Samek-Lodovici relates it to a more general constraint against structure (*STRUC, Prince and Smolensky [1993] 2004), but NOFEATS is not just a constraint against features (otherwise the noun would violate it too), but is a constraint against features that are the result of a relationship (agreement). The locus of potential violations (the modifiers) is different from the locus of exceptionality (the noun). So, it would be tricky to rule out candidates with features that are not the result of agreement, but that have been provided by GEN.

The facts of Spanish are not easily accounted for in A&N's model either. The loss of [FEMININE] in prenominal position would be the result of a weakening rule applying within a prosodic phrase in the context of a feminine noun in the same prosodic domain. Leaving aside for a moment the restriction to specific lexical items, the weakening rule could have the form in (19), where X stands for any prenominal categories that can coappear with a feminine noun.

(19) {[X FEM SG . . .] [N FEM SG . . .]} → {[X SG . . .] [N FEM SG . . .]}

This rule would cause the loss of [FEMININE] in singular elements in prenominal position but it would not affect modifiers in postnominal position. Plural nouns would not be affected by (19). In this case, again, the trickiest part would be how to restrict the rule to apply only to specific lexical items, especially given that A&N assume late insertion, and that the spell-out of terminals applies after the application of context-sensitive allomorphy rules. One possibility is to assume, with Embick and Halle 2005 and other work, that roots are phonologically present already in the syntax or, alternatively, that they can ultimately be identified through abstract indices or some sort of diacritic feature.[9] Then the rule in (19) could be slightly modified as in (20) to affect only specified nouns. (20a) and (20b) constitute two ways of notating the restriction to specific lexical items.

(20) {[X FEM SG . . .] [N* FEM SG . . .]} → {[X SG . . .] [N* FEM SG . . .]}
      a. N*: *arma, agua, área,* . . .
      b. N*: N^{22}, N^{317}, N^{683}, . . . (where N^{22} = *arma*, N^{317} = *agua*, N^{683} = *área*, . . .)

10.4   A Challenge for A&N 2003: Gender and Mass Concord in Asturian

Asturian nominals have, in addition to gender and number, a distinction between mass and count.[10] For instance, the masculine noun for 'thread' is realized as *filu*, with the ending *-u*, when it is interpreted as a count noun, and as *filo*, with the ending *-o*, when it is interpreted as a mass noun. Postnominal adjectives show the same *-u/-o* distinction, as illustrated in (21). Feminine nouns, like *cebolla* 'onion', have the ending *-a* under the two interpretations, but the count-mass distinction does surface in postnominal adjectives, where the ending for mass interpretation is *-o*, the same one that is found with masculine nouns. This appears illustrated in (22).

(21) a. fi**lu** blan**cu**   MASC, COUNT
      b. fi**lo** blan**co**   MASC, MASS
         'white thread'

(22) a. cebol**la** blan**ca**   FEM, COUNT
      b. cebol**la** blan**co**   FEM, MASS
         'white onion'

However, prenominal elements agree with the noun only with respect to gender, not with respect to mass. The examples in (23) and (24) show the asymmetries in concord when the noun receives a mass interpretation. The noun *fierru/o* in (23) means 'iron', and *manzana* in (24) means 'apple'; the relative order of the adjectives is the same as in English.

(23) a. du**ru** fie**rru** ferruñosu   MASC, COUNT
    b. du**ru** fie**rro** ferruñoso   MASC, MASS
       'hard rusty iron'

(24) a. gua**pa** manzana madu**ra**   FEM, COUNT
    b. gua**pa** manzana madu**ro**   FEM, MASS
       'good ripe apple'

In (23b) and (24b) the noun has gender and mass features; gender spreads to prenominal elements while mass spreads to postnominal elements. (25) shows schematically how gender and mass concord work in Asturian. The asymmetries arise only when the noun has a mass interpretation, as (25a) illustrates. When the noun has a default count interpretation, gender spreads to prenominal and postnominal elements, as shown in (25b).

(25) a. Mass interpretation

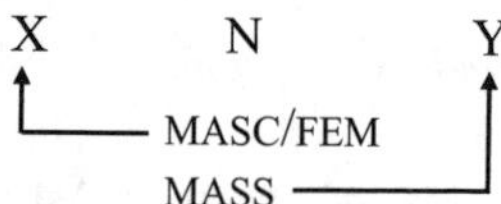

b. Count interpretation

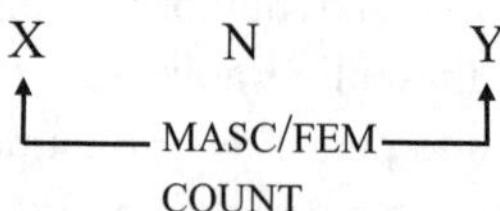

Asturian concord asymmetries cannot be accounted for in a straightforward fashion through weakening rules of the type proposed in A&N 2003. Given that the structure of the DP looks, in all relevant respects, identical to the one for Spanish, the initial prosodic structure that creates the domains for the weakening rules would be the same: {X N} {Y}. Syntactically full concord for gender and mass would affect the whole DP, which at that point contains only abstract terminals. A weakening rule deleting all mass features but the last within the first prosodic phrase {X N} would account for the fact that only gender concord surfaces prenominally. But the preference for the exponence of mass over the exponence of gender for all postnominal elements cannot be accounted for by a weakening rule; in that domain, {Y}, mass and gender are wrongly predicted to survive: the Recoverability condition does not allow deletion because there is no other terminal with those features within that prosodic domain. It could be argued that, because the adjectival endings realize either gender or mass (masculine: *-u*, feminine: *-a*, mass: *-o*), some additional mechanism favors the realization of mass over gender. But this mechanism would be very different from the one that applies prenominally, even though the conflict between exponents would also be present in that position. Resorting to other prosodic phrasings would not improve the analysis. If the phrasing were {X} {N Y} (with left alignment in (6), instead of right alignment), the problems would be transfered to prenominal elements. The only phrasing that would allow for an account within this model would be to posit a single prosodic phrase for the whole DP: {X N Y}. One weakening rule, (26a), would delete the mass feature from prenominal elements and another weakening rule, (26b), would delete the gender feature from postnominal elements.

(26) a. {[X ±FEM MASS . . .] [N ±FEM MASS . . .]} → {[X ±FEM . . .]
    [N ±FEM MASS . . .]}

  b. {[N ±FEM MASS . . .] [X ±FEM MASS . . .]} → {[N ±FEM MASS . . .]
    [X MASS . . .]}

The relevant question is, though, how this initial prosodic phrasing could be obtained. It would have to be assumed that syntactically only the head N, instead of the NP, raises when N is not final, contrary to what one would have to assume for Spanish, a very closely related language. This relates to a more general problem, mentioned in section 10.2: to prevent more than one prosodic phrase from being created within the DP, no right (or left) XP boundaries should exist internally to the DP, a very strong claim that goes in the opposite direction from most analyses of DP structure (see, among many others, Svenonius 2008).[11]

For Samek-Lodovici's model, which does not resort to prosodic phrasing, the data from Asturian can be accounted for with no modification of the model. A top-ranked AGR$_{mass}$ constraint would favor postnominal mass concord, and the lower-ranking EXTAGR$_{gen}$ >> EXTAGR$_{mass}$ would force gender concord prenominally.

10.5 Concord in Two Steps

The model to be sketched in this section shares with A&N the idea that agreement asymmetries are obtained in two steps, a syntactic one and a postsyntactic one. With Samek-Lodovici it shares the optimality-theoretic orientation, with some very similar constraints. A more detailed description of this model, with examples from several languages, can be found in Bonet, Lloret, and Mascaró (BL&M), forthcoming, and in Bonet and Mascaró (B&M) 2012.

In the BL&M model, contrary to A&N, the syntax is responsible only for postnominal concord, most likely through spec-head agreement. This assumption is compatible with claims in Shlonsky 2004 or Nevins 2011a, mentioned in section 10.2, and would cover a similar agreement domain as the markedness constraint AGR$_f$ in Samek-Lodovici 2002 as applied to the DP in sections 10.2 and 10.3. Postsyntactically, faithfulness constraints of the MAX family (27a), which protect postnominal concord, and general CONC(ORD) markedness constraints, (27b), compete with each other and are the source of the existing asymmetries. CONC(ORD) is equivalent to Samek-Lodovici's EXTAGR$_f$. The scope of the two types of postsyntactic constraints is represented in (28).

(27) a. MAX: Every inflectional feature of the input has a
                  correspondent in the output.

  b. CONC(ORD): If a N has an inflectional feature F, all other modifiers
                  within the DP must have the inflectional feature F.

(28) X   N   Y

                MAX

          CONC

This model, like Samek-Lodovici's, can account in a straightforward fashion for the concord asymmetries found in Asturian. The tableau in (29) shows schematically how the asymmetric concord in a sequence like *guapa manzana maduro* 'good ripe apple', (24b), would be obtained.

(29)

Input: X N Y			MAX (mss)	CONC (gen)	CONC (mass)	MAX (gen)
	FEM.MSS	FEM.MSS				
a. ☞ X N Y				*	*	*
FEM	FEM.MSS	MSS				
b. X N Y			*!		**	
FEM	FEM.MSS	FEM				
c. X N Y				**!		*
MSS	FEM.MSS	MSS				
d. X N Y			*!	*	*	
MSS	FEM.MSS	FEM				

The modifiers can realize only mass or gender (typically -*o* for mass, -*u* for masculine, -*a* for feminine), and these morphs are introduced by GEN. For this reason, no candidates are considered in (29) with modifiers that express both gender and mass.

This analysis overcomes the problems mentioned in section 10.4 for the approach by A&N, which relies crucially on initial prosodic phrases. The BL&M proposal also avoids some of the difficulties the one-step approach by Samek-Lodovici runs into for the concord asymmetry found in dialects of Spanish with lexical items like *agua* or *arma*. Recall that only certain feminine nouns cause the asymmetry, and that the asymmetry arises only in the singular; when the noun is plural concord applies in a totally regular fashion. Postnominal gender and number concord takes place in the syntax. Postsyntactically an impoverishment rule deletes the feature [feminine] in a specified set of nouns, as shown in (30).

(30) [FEM] → Ø / __[SG] for *agua, arma, ave, . . .*

The loss of the feature [feminine] prevents it from spreading to prenominal elements. The tableaux in (31) and (32) show very schematically how lazy concord is obtained in the singular but not in the plural, with full concord. The constraint CONC(FEM.PL) is a shorthand for two constraints that favor concord for marked gender and number features—that is, feminine and plural. This constraint is not relevant in (31) because the N has neither feminine nor plural features, but it becomes crucial in (32), where it prevents the concord asymmetry. The constraint *FEM(ININE) belongs to the family *STRUC, against structure.

(31) Lazy concord with singular nouns like *agua* 'water'

Input: X N Y _SG FEM.SG	MAX	CONC (FEM.PL)	*FEM
a. ☞ X N Y SG _SG FEM.SG			*
b. X N Y FEM.SG _SG FEM.SG			**!
c. X N Y SG _SG SG	*!		

(32) Full concord with plural nouns like *aguas* 'waters'

Input: X N Y FEM.PL FEM.PL	MAX	CONC (FEM.PL)	*FEM
a. X N Y PL FEM.PL FEM.PL		*!	**
b. ☞ X N Y FEM.PL F.PL FEM.PL			***!
c. X N Y PL FEM.PL PL	*	**!	*

10.6   Conclusions

This chapter has compared three possible analyses of DP-internal concord asymmetries, the first two, Samek-Lodovici 2002 and A&N 2003, having originally been designed to handle subject-verb asymmetries. The one-step analysis by Samek-Lodovici would especially run into problems with the fact that, in certain dialects of Spanish, the asymmetry is triggered by specific nouns and only in the singular. In A&N's proposal, a two-step process, there is full concord syntactically and, at PF, weakening rules restricted to prosodic phrases determine the final asymmetric concord. This proposal raises several questions concerning the construction of initial prosodic phrasing and encounters problems especially with the analysis of Asturian, where there is gender concord prenominally but mass concord postnominally. In the two-step proposal advocated here, postnominal agreement takes place syntactically and is controlled postsyntactically by faithfulness constraints, which confict with a general agreement markedness constraint. This proposal does not encounter the problems faced by the other proposals in the analysis of the Spanish and Asturian facts.

In the very sketchy analysis of Spanish and Asturian there are issues that have not been addressed and that will be left open here due to space considerations. An important one has to do with the interaction between asymmetries and late insertion. This issue was briefly addressed in the last part of section 10.3, concerning the A&N model, which also assumes late insertion. To restrict lazy concord in Spanish to specific nouns, it was suggested that, following Embick and Halle 2005, it could be assumed that roots or abstract indices (or some diacritic feature) identifying them could be present in the syntax. Another issue is the timing and manner of exponence of the impoverished noun in Spanish. An item like *agua* has the same class marker *-a* in the singular, where it has lost the feature [feminine], and in the plural (*aguas*), where it has kept it. This would not be a problem for proposals like A&N, where the feature makeup of the Noun does not change during the derivation. There are several ways to approach this issue within the present model. To mention one of them, the final *-a* could be considered a stem formative in the sense of Bermúdez-Otero 2006, rather than a gender or class marker. Then the only exponent possible for 'water' would be *agu-a*, regardless of the presence or absence of the feature [FEMININE].

Notes

I would like to thank Ora Matushansky and an anonymous reviewer for their helpful comments. Thanks also go to Joan Mascaró and Carme Picallo for comments and discussion on the first draft of this chapter, parts of which were presented at the University of Manchester (February 2012) and at the Workshop on the Selection and Representation of Morphological Exponents (IS/CASTL, Tromsø, June 2012). I thank the audiences at these two events for their comments. This research received financial support from the Spanish Ministry of Education (grant FFI 2010-22181-CO3-01) and the Generalitat de Catalunya (research group 2009SGR1079).

1. A different issue is the competition that arises in languages like Georgian where both the subject and the object trigger verb agreement. See McGinnis (chapter 3, this volume) for a proposal that accounts for the lack of double number marking on the verb.

2. An anonymous reviewer points out that there is a clear configurational difference between clausal agreement and DP agreement: in clausal agreement the subject is an XP while in DP agreement the noun is a head. Nevertheless, several authors (most extensively, Cinque 2010) have argued for NP movement and against N movement within the DP. As briefly described in section 10.2, Shlonsky (2004) explicitly argues that not only movement but also concord within the DP is achieved with the NP, not N.

3. There are other proposals on DP–internal agreement that do not consider the asymmetries that can arise between prenominal and postnominal modifiers (see, for instance, Picallo 2008 or Schoorlemmer 2009). It is unclear how they would account for these asymmetries.

4. Number and gender features appear in this chapter with the notation typically attributed to privative features (e.g., [PLURAL]) to simplify their representation. See Harbour (chapter 8, this volume) for arguments in favor of binary number features.

5. A&N (2003) use the term *prosodic phrase* to refer to some intermediate structure, between syntactic structure and real prosodic structure. The structure cannot be strictly prosodic because at that point the insertion of phonological material has not taken place yet, and therefore the phrasing cannot be sensitive to weight, for instance.

6. For concreteness and clarity, in (15) I assume that the input contains phonologically realized roots plus a noun with realized inflection. It is not clear whether Samek-Lodovici would endorse late insertion.

7. Most proponents of lexically indexed constraints defend the view that these should be restricted to faithfulness constraints. This is an unavailable option here because all the constraints under consideration are markedness constraints.

8. The presence of full concord with plural nouns is a historical residue of the older restriction against a contact between low vowels. In the plural, the presence of the plural morph -*s* broke the adjacency between the two vowels.

9. Abstract indices are posited in Embick and Halle 2005 to differentiate homophonous roots.

10. Due to limitations of space, number is ignored in what follows. Number is incompatible with mass. See Bonet and Mascaró 2012 for a more detailed description and analysis of Asturian concord asymmetries; see also Mascaró 2011.

11. An additional problem for A&N's model is related to the idea that asymmetries in agreement are obtained before Vocabulary Insertion. Evidence that this cannot be true comes from North-Eastern Central Catalan (NEC Catalan), where plural concord fails

to surface only prenominally, and crucially only when the plural morph *-s* would surface between consonants, as the following examples show (see Bonet, Lloret and Mascaró, forthcoming, for more examples and argumentation, see also Nevins 2011a). The agreement asymmetries must arise at least at the same time Vocabulary Insertion takes place.

(i) a. *molt_ poc_   professional-s bon-s   present-s*
       much few   professional.PL good.PL present.PL
       'very few present good professionals'
   b. *el-s   antic-s amics*
       the.PL old.PL friend.PL
       'the old friends'

11 Suspension across Domains

Jonathan David Bobaljik
Susi Wurmbrand

11.1 Shifting Domains

The notion of a cyclic derivation, defining (sub)domains in a grammatical
derivation to which rules apply, dates from some of the earliest work in modern
linguistics and is a recurring theme in the work of Morris Halle (see, famously,
Chomsky and Halle 1968). In phonology, it is recognized that not all morpheme concatenation triggers cyclic rule application, but that certain morphemes are designated triggers of cyclic rules (see, e.g., Halle and Vergnaud
1987 for one approach). A related idea pervades the history of syntax, holding
that there are cyclic domains defined (at least in part) with reference to particular heads/projections, and that these cycles enforce locality conditions on
syntactic dependencies (see, e.g., Ross 1967; Chomsky 1973, 1986). Within
the intermodular perspective of Distributed Morphology (Halle and Marantz
1993) various authors have raised the question of how the domains (e.g.,
cycles, phases) of one module (syntax, morphology, semantics) interact with
those of others (see also Scheer 2008 and related work).

In this short chapter, we explore one small aspect of this large puzzle. Specifically, we propose a general rubric that allows for some slippage in otherwise well-established locality domains—cases in which a well-motivated
cyclic domain appears to be suspended, allowing dependencies to span a larger
structure than they normally may. To the extent that this is on the right track,
it bolsters arguments that cyclic domains constrain the locality of operations
across modules and thus constitute a deep property of grammar. Specifically,
we suggest that the following *Domain Suspension* principle holds across
modules, and present two applications, one from suppletion in morphology,
the other from quantifier raising (QR), suggesting the potentially broad applicability of the principle.[1]

(1) In the following configuration (linear order irrelevant), where the
      projection of Y would normally close off a domain, formation of such a
      domain is *suspended* just in case Y depends on X for its interpretation.
      [X [$_Y$^{n} Y]]

Although (1) could be implemented in various ways, we conceive of suspension not as an operation, but as a condition restricting (or defining exceptions
to) the algorithm(s) that determine(s) derivationally whether a given maximal
projection will or will not constitute (or close off) a domain. Various terms in
this general scheme (notably *domain, interpretation*) are relativized to some
extent, to the module under consideration, accounting for a slight difference
in the ways in which (1) plays out in the different components of grammar.
For the cases considered here, the algorithms subject to Domain Suspension
in the structure in (1) include:

(2) a. *Morphology*: If X is a cyclic head, then Y^n is a spell-out domain,
          unless Y depends on X for its interpretation.
      b. *Syntax*: If Y^n is the highest projection of a (potential) cyclic domain,
          then Y^n constitutes a phase, unless Y depends on X for its
          interpretation.

We illustrate these in turn.

11.2   *Optimal* Suspension: Superlative Suppletion

The first case of (1) that we consider is in the morphology, and is drawn from
the study of adjectival suppletion in Bobaljik 2012. We limit ourselves to a
brief presentation here and refer the reader to the work cited for additional
detail and important qualifications.

   In a comprehensive survey of suppletion in adjectival gradation (*good–
better–best*), encompassing some 116 distinct suppletive cognate triples (POSI-
TIVE–COMPARATIVE–SUPERLATIVE) from more than 70 languages, Bobaljik
reports the following patterns.

(3) Regular   A   A   A   *big–bigger–biggest*
      Suppletive   A   B   B   *good–better–best*
      Doubly suppletive   A   B   C   *bonus–melior–optimus*
      Unattested   A   B   A   **good–better–goodest*
                                   A   A   B   **good–gooder–best*

Bobaljik argues that the (relative) superlative is universally built from the
comparative, hence the underlying structure is (hierarchically) [[[ROOT] CMPR]
SPRL]. The structure is transparent in many languages, illustrated in (4), but

obscured by a null comparative allomorph in the superlative in some (such as English). That is, we assume that even in English, the superlative has the structure [[[big] Ø_{ER}] est] (Bobaljik 2012 constitutes an extended defense of this analysis; see also Stateva 2002).[2]

(4)		POS	CMPR	SPRL	GLOSS
a.	Persian	bozorg	bozorg-tar	bozorg-tar-in	'big'
b.	Cimbrian	šüa	šüan-ar	šüan-ar-ste	'pretty'
c.	Czech:	mlad-ý	mlad-ší	nej-mlad-ší	'young'
d.	Hungarian:	nagy	nagy-obb	leg-nagy-obb	'big'
e.	Latvian:	zil-ais	zil-âk-ais	vis-zil-âk-ais	'gray'
f.	Ubykh:	nüs°ə	ç'a-nüs°ə	a-ç'a-nüs°ə	'pretty'

This nested structure, along with the assumption that suppletion is the result of contextual allomorphy of the root, as in (5), explains the absence of the ABA pattern: in the absence of a more specific rule, the Subset Principle (Halle 1997a) ensures that (5a) will bleed (5b) in the comparative and in the superlative (because the environment for application is met in both instances).

(5) a. GOOD → *be(tt)* / __] CMPR
    b. GOOD → *good* /   <elsewhere>

A simple explanation for the complete absence of the AAB pattern, which we assume is essentially correct, supplements the account just given with the additional assumption that the CMPR node is cyclic (i.e., domain-defining), and that the SPRL morpheme is thus insufficiently local to be able to serve as the context triggering a suppletive root (see Embick 2010 and Bobaljik 2012 for different views on the interaction of cyclicity and locality for allomorphy).[3] Recall that, by hypothesis, the superlative is derived from the comparative. At the intermediate point in this derivation where the comparative head is introduced, the representation is as in (6a).

(6) a. [[√ ROOT] CMPR]
    b. [[[√ ROOT] CMPR] SPRL]

We assume CMPR is cyclic, and thus it triggers spell-out of its complement. That is, the complement of CMPR is a spell-out domain, which we take to mean, at a minimum, that rules of exponence/Vocabulary Insertion apply at this stage. For an adjective such as *big* that lacks a suppletive comparative allomorph along the lines of (5a), spell-out will apply and insert the default form of the root. When the SPRL morpheme is subsequently added (as in (6b)), it is too late to trigger a special rule at the root cycle, since the form of the root has already been fixed.

Attractive as it may seem (especially in that the *AAB pattern is robustly unattested, constituting zero of the 116 suppletive triples in Bobaljik's sample), it is factually incorrect that the superlative cannot govern a suppletive root allomorph. The Latin ABC triple *bonus–melior–optimus* 'good–better–best' shows this to be the case (there are a handful of other such patterns, including in Welsh and Middle Persian). The generalization appears to be this: the superlative head is accessible as a context for root allomorphy just in case the comparative also triggers root allomorphy (ABC), but if the comparative is regular, then so too must the superlative be regular: (AAX → AAA). Domain Suspension in (1) explains why this is so, as follows.

We take (7) to be the representation of the Latin 'good' root allomorphs.[4]

(7) a. GOOD, CMPR   → *opt-*   / __] SPRL
     b. GOOD   → *mel-*   / __] CMPR
     c. GOOD   → *bon-* /   <elsewhere>

A root like big that does not have a comparative allomorph does not depend on the node CMPR for its interpretation, that is, for the choice of exponent. This allows such roots to be spelled out on the cycle introduced by CMPR as indicated above. By contrast, the Latin (7) and English (5) roots meaning 'good' *do* depend on the head CMPR—the exponent to be selected in a derivation cannot be determined without reference to this head. As such, Domain Suspension is triggered and spell-out is delayed (as it were) until the next higher cycle. At the next higher cycle, the head SPRL is introduced, and is thus available to trigger allomorphy (perhaps subject to an adjacency requirement, and thus permitting only portmanteau allomorphy for the root plus CMPR as in (7a)).

Note, importantly, that Domain Suspension, as formulated, does not permit of a hypothetical Latin′, just like (7) but omitting (7b), and yielding the unattested AAB pattern: **bonus–bonior–optimus*. Establishment of a spell-out domain is only suspended if the head of the erstwhile domain depends on the cyclic head for its interpretation. A rule like (7b) thus establishes exactly the right type of dependency to suspend the domain, but a rule like (7a), which only indirectly affects the form of the adjectival root, but does not operate on the root node as such, does not.

In sum, the idea that locality is defined (at least in part) in terms of (cyclic) domains, provides, in the case at hand, a straightforward account of an exceptionless crosslinguistic generalization, namely, the absence of *AAB patterns in adjectival suppletion. Patterns of this sort simply do not exist. On the other hand, the Domain Suspension Principle in (1) provides a loophole, allowing for a very narrow class of cases in which locality is weakened and where suppletion of (a constituent including) the root may be governed by the superlative. Domain Suspension requires, correctly, that these cases have certain hallmark properties—at a minimum, they must constitute ABC patterns, in which the comparative is independently suppletive, since it is exactly the rule that yields comparative suppletion that prevents the establishment of an opaque (i.e., cyclic) domain including the root.

11.3   Tense Dependencies: Phase Suspension and QR

We now explore the analog of (1) in syntax. In this component, we take phases to be the crucial locality domain regulating the locality of various dependencies. It is well established that infinitives and subjunctives, in contrast to finite indicative clauses, are transparent for various properties crosslinguistically (e.g., long-distance reflexive binding, Condition B transparency, NPI-licensing, Case licensing, A-movement, control, scope, and others). The degree of porosity of the embedded clauses often goes along the following continuum: *finite » subjunctive » infinitive » raising*. QR in English poses an odd puzzle: it is possible from control and ECM infinitives, as well as subjunctive clauses, but not from finite clauses (at one end of this continuum), nor from raising constructions, which are often assumed to be the most transparent configurations. This is illustrated by the examples in (8) (see Kayne 1981, 1998; Longobardi 1992; Lebeaux 1995; Bayer 1996; Kennedy 1997; Fox 1999, 2000; Johnson 2000, among others, for further examples, qualifications, and controls).

(8) a.   #Someone said that Sue is married to every man.

$\forall » \exists$ (finite)

  b.   She has requested that they read only Aspects.

only » request (subjunctive)

  c.   A different student decided to report on every article.

$\forall » \exists$ (infinitive)

  d.   Someone expects Sue to marry every boy.

$\forall » \exists$ (infinitive)

  e.   #This soldier seems to someone to be likely to die in every battle.

*$\forall » \exists$ (raising)

Wurmbrand (forthcoming) adopts a cyclic spell-out model according to which completed cycles (phases) are subject to transfer followed by (LF and PF) spell-out of the complement of the phase head. Since, per common assumption, a *spelled-out* domain is inaccessible for further syntactic operations (such as movement), this model has as a consequence that movement (whether overt or covert) is phase-bound (see also Cecchetto 2003, 2004; Takahashi 2010). The main proposal regarding the distribution of QR in (8) is that finite clauses and raising infinitives involve a solid phasal domain, hence block QR, whereas control and ECM infinitives trigger domain suspension in (1), hence allow QR.

A simplified structure for finite and raising complements is given in (9).[5] In finite embedded complements, XP corresponds to CP, in raising infinitives to AspP (see Wurmbrand 2011). Crucially, in both constructions, we argue, XP constitutes a phase. Following ideas in Bobaljik and Wurmbrand 2005 and Bošković 2010, we pursue a dynamic phasehood view; specifically we propose that the highest projection of a clause, no matter what category or label, constitutes a phase. This is responsible for making the CP of a finite clause as well as the AspP of a raising infinitive a phase, hence a locality domain for movement.

(9)

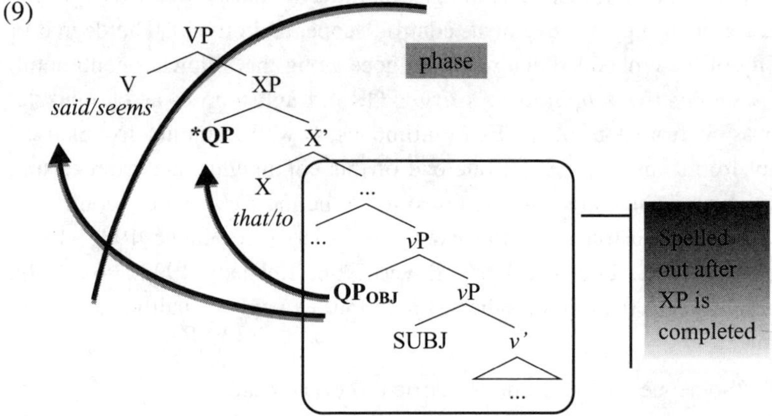

Before laying out how exactly phasehood is determined in this dynamic phasehood model, consider the effects of the assumption that XP in (9) is a phase. First, a derivation that involves movement of a QP from the complement of the phase head X across XP is excluded. Since the complement of X is a spell-out domain, any material within that domain will be inaccessible for further operations. Second, a derivation involving successive cyclic movement of a QP to Spec,XP is excluded by Fox's (2000) *Scope Economy* condition. Scope Economy is essentially a Last Resort condition for covert movement that states that each step of QR must be semantically motivated and cannot be semantically vacuous. Since movement of a QP to Spec,XP does not yield a different interpretation, it is excluded by Scope Economy. For an embedded QP to take matrix scope, however, movement through Spec,XP would be necessary, since otherwise the QP is trapped in a spell-out domain. But since successive cyclic movement is not available, QR to the matrix clause is excluded. Thus, the dynamic phase view proposed here unifies the clause-boundedness effects found in English finite indicative clauses and raising infinitives.

While phasehood of finite CPs is uncontroversial, the claim that raising infinitives are phases is less obvious. One piece of motivation is provided by the binding properties in (10) (due to Fox, cited in Grohmann, Drury, and Castillo 2000; Bošković 2002; Pesetsky and Torrego 2007; Castillo, Drury, and Grohmann 2009, among others). If the top projection of a raising infinitive (typically assumed to be TP, but see Wurmbrand 2011 for reasons to assume it is an aspectual projection) is a phase, movement of the subject always has to proceed through the specifier of the top projection, AspP in (10). The existence of a copy of the subject in Spec,AspP then correctly accounts for the fact that the anaphors in these examples always have to be bound by the subject rather than the higher experiencer.[6]

(10) a.   [John seems to Mary [$_{\mathrm{AspP}}$ ~~John~~ to appear to himself
            [$_{vP}$ ~~John~~ to be. . .]]]

    b. *[Mary seems to John [$_{\mathrm{AspP}}$ ~~Mary~~ to appear to himself
            [$_{vP}$ ~~Mary~~ to be . . .]]]

Assuming the dynamic phase approach above, the obvious question now is why control and ECM infinitives, as well as finite subjunctive complements, are not phases—that is, why phasehood is suspended in these cases but not in finite complements and raising infinitives. As stated in (1), we propose that domain establishment is suspended just in case the lower head (in the cases at hand, the top head of the embedded clause) depends on the head that combines with it for its interpretation. To flesh this out, we must first return to a point put aside above, namely the question of how phasehood is determined in a dynamic phase model in general.

In contrast to Chomsky 2000, 2001, where certain designated projections—only (strong) vP and CP (maybe also DP, PP)—are phases, we argue that it is particular domains that determine phases, and that it is the highest projection of these domains (again whatever their category or size) that normally constitutes a phase. These potential cyclic domains (i.e., potential phases) are specified as follows (see Wurmbrand 2012 for evidence for this view from ellipsis, among others):[7]

(11) a. Aspect domain: Theta-domain plus any event structure/Aktionsart-
            dependent aspect (progressive, perfective, imperfective)
    b. T+C-domain: Discourse domain, mood, tense, modal domain

Under this approach, no phrase is inherently a phase and the lack of a CP or vP does not entail the lack of phase (i.e., unaccusative and passive VPs also qualify as phases, which is motivated by the binding and reconstruction properties discussed in Legate 2003 and Sauerland 2003). How then does a syntactic object "know" whether it is a phase or not? We propose that phasehood is determined strictly derivationally: when an XP merges with a head Y, XP becomes a phase if Y is part of the next (potential) cyclic domain (e.g., if vP merges with T, vP becomes a phase). As a result, XP is subject to Transfer at this point. On the other hand, if XP merges with a head Y that is part of the same (potential) cyclic domain as XP (e.g., if vP merges with a progressive head), XP does not become a phase but its cyclic domain is extended to YP (and potentially further) and Transfer does not occur yet.

With this system in place, we are now in a position to tackle domain suspension in control/ECM and subjunctive clauses. The crucial idea is that merging a control, ECM, or subjunctive verb with its complement involves a particular semantic dependency, specifically, a featural dependency spanning the domain boundary, which suspends phasehood of the complement. Although the complement constitutes a separate potential cyclic domain in the sense of (11), the dependency in question blocks the establishment of an actual domain of cyclic rule application—a syntactic phase—in these cases, with consequences for the locality of movement.

To see what kind of dependency this is, a short detour to the tense properties of infinitives is necessary. In addition to raising infinitives, which, we argue, involve aspect but no tense, there are two types of infinitives in English that differ regarding their tense properties: future irrealis infinitives and simultaneous propositional infinitives. In Wurmbrand 2011, it is shown that both types of infinitives can be realized as control and ECM constructions, and it is proposed that future irrealis infinitives involve an abstract future modal *woll* but no tense. By contrast, simultaneous propositional infinitives involve a *zero* tense, representing the attitude's holder "now" (Kratzer 1998; Abusch 2004). The major evidence comes from sequence of tense phenomena (Ogihara 1995, 1996, 2007; Abusch 1997; Enç 2004) and the distribution of eventive predicates—that is, nonstative, nongeneric, episodic predicates (Pesetsky 1992; Bošković 1996, 1997; Martin 1996, 2001). We briefly illustrate the latter property. As shown in (12), future infinitives, like finite future contexts, allow eventive predicates, whereas simultaneous propositional infinitives, like present-tense contexts, prohibit eventive predicates (see Abusch 2004 and Wurmbrand 2011 for several further examples).

(12) a. Leo decided/plans to bring the toys tomorrow.   Control
     b. The printer is expected to work again tomorrow.   ECM

    c. Leo will leave tomorrow.   Finite future
    d. Yesterday, John claimed to *leave/√be leaving (right then).   Control
    e. Yesterday, John believed Mary to *leave/√be leaving (right then). ECM
    f. Leo *sings/√is singing right now.   Present

Wurmbrand (2011) proposes that the distribution of eventive predicates is an effect of aspect. Specifically, it is argued that present tense and zero tense are only compatible with imperfective aspect, which, in English, corresponds to the *-ing* form. Perfective aspect (which is not realized morphologically in English), requires that the event time interval be included in the reference time interval, which is not possible in present- and zero-tense contexts, since the reference time interval (the utterance time or the attitude holder's now) is too short. Future contexts, on the other hand, involve the modal *woll*, which extends the reference time, and hence allows a perfective interpretation—an interpretation where the event time interval is included in the reference time interval. Assuming that propositional infinitives involve a zero tense and future infinitives a future modal *woll* thus allows a unified account of the distribution of eventive predicates based on the interaction of tense and aspect.

The tense properties of control and ECM infinitives play a crucial role in our account of domain suspension in these contexts (as well as subjunctive complements) in English. Comparing the tense/mood/aspect properties of the different types of embedded structures, one clear difference arises: in control/ECM infinitives and subjunctive complements, the specific semantic value of the highest head is selected by the matrix verb (as part of its lexical properties). In finite clauses and raising infinitives, on the other hand, no such selectional relation exists.

The examples in (13) illustrate the dependency between specific matrix predicates and the temporal/mood composition of the complement: *demand* (vs. *say*) selects subjunctive, *decide* selects a future infinitive, and *claim* (also *believe*) selects a simultaneous infinitive. Finite complements of *say*, on the other hand, show no tense restriction imposed by the higher verb, as in (14a). Similarly, aspect in raising infinitives (which are semantically tenseless) is not a property of the verb *seem* but a combination of the type of embedded predicate (stative vs. nonstative) and the higher tense, as in (14b).

(13) a. I demand/*said that he listen to this.
     b. Mary decided to leave tomorrow/to become/get/#to be pregnant.
        (only if future state)
     c. Mary claimed to {be/*become/*get} pregnant/*to leave tomorrow.

(14) a. John said that he ate/is eating/will eat a cookie.
    b. John seems to be sleeping/*to sleep right now/to sleep whenever
        he's tired.

More abstractly, we assume that the dependencies in (13) are dependencies between an underspecified (or unvalued) feature of the potential phase head and a lexical value of the verb the clause combines with, whereas there is no feature dependency in (14). The schemas here illustrate the relevant portions of the corresponding examples above:

(13′) a. *demand*$_{SBJV}$ [$_{XP=potential\ phase}$ $X_{F:}$ ___   . . .]
     b. *decide*$_{FUT.IRR}$ [$_{XP=potential\ phase}$ $X_{F:}$ ___   . . .]
     c. *claim*$_{SIM}$ [$_{XP=potential\ phase}$ $X_{F:}$ ___   . . .]

(14′) a. *say*   [$_{XP=phase}$ $X_{F:\ val}$ . . .]
     b. *seem*   [$_{XP=phase}$ $X_{F:\ val}$ . . .]

It is this feature-selection dependency, we argue, that suspends phasehood and thus also spell-out.[8] The feature values under consideration are all values that need to be visible in semantics to properly interpret the structure. Thus, in jargon, the features that show a value dependency in (13) are all interpretable features. The intuition behind (1) is then that a (potential) cyclic domain that is incomplete in a crucial semantic way (i.e., the topmost head is semantically underspecified and only interpretable in conjunction with the selecting head) cannot be transferred, and spell-out is suspended.

A specific technical implementation of value selection is given in Wurmbrand (forthcoming). It is assumed there that verbs that impose a value-selection restriction on their complements are lexically specified with an uninterpretable valued feature encoding the specific value. For instance *decide*, *expect* are specified for *u*F: *woll*, whereas *claim*, *believe* are specified for *u*F: $\emptyset_T$, and subjunctive-taking verbs like *demand* are specified for *u*F: *subjunctive*. The topmost head of the complement, on the other hand, is underspecified in that it comes unvalued. Crucially, since those heads encode semantic information (tense, mood, modality), these features are still interpretable features. If an interpretable feature is sent to LF without a value, LF could not assign an interpretation, and the structure would not be interpretable. Thus, the only way the structure will converge is if the unvalued features are valued via Agree before LF.

The features as specified above have the effect that a mutual dependency is established between certain verbs and corresponding types of complements. The unvalued features of the top clausal projection need to enter an Agree relation with a higher verb that has an *u*F: val. Similarly, the uninterpretable

feature of the selecting verb also becomes dependent on a specific complement, as desired. Following Pesetsky and Torrego 2007, uninterpretable features (whether valued or unvalued) need to be licensed; specifically, they need
to be connected to a corresponding interpretable feature (cf. the proposal of
the *Thesis of Radical Interpretability*, Brody 1997).

The feature specification proposed allows us to address the question of why
complements that are value-selected are not phases. There are two ways to
implement this, and we will leave the choice between the two options open
here. First, it could be assumed that valuation suspends phasehood. Since the
unvalued features under consideration are interpretable features, these units
would be interpretationally incomplete (before valuation takes place), and
hence at the point where the clauses are completed, these units would not
qualify as objects that are usable by the semantics. Alternatively, it could be
assumed that the heads with the unvalued features undergo head movement,
which causes phase extension (see Den Dikken 2007) or phase sliding (see
Gallego 2005, 2010; Gallego and Uriagereka 2006), although note that these
latter proposals, unlike Domain Suspension as we have defined it, do not obviously extend to the morphological cases discussed in section 11.2.

11.4 Conclusions

In (13), the head X head is deficient, in the sense that it depends on the next
head up for its (temporal) interpretation. Empirically, this is manifested as a
type of selection relation—verbs like *demand* and *decide* require a particular
temporal interpretation of their (infinitival) complement, where verbs selecting
raising and finite complements do not. This dependency makes subjunctive,
control, and ECM complements special. One aspect of their special nature is
that they fail to constitute a phase (domain), and thus do not restrict movement
in the way phases do—they do not require an element moving across them to
land in their periphery. QR has the perfect properties to show this effect. Since
(long) QR is not driven by feature-checking needs, it is subject to Scope
Economy, and thus, effectively, cannot escape a phase. Long QR is thus only
possible where no phase intervenes, and serves as an excellent phase
detector.

In (5) and (7), the adjectival root is deficient, in the sense that it depends
on the next head up for its (phonological) interpretation. Empirically, this is
manifested as a selection relation—for suppletive adjectives, the comparative
requires a particular allomorph of the root, where for nonsuppletive adjectives,
no such selection relation obtains. This dependency (suppletion) makes roots
with comparative allomorphs special—one aspect of their special nature being

that they fail to be "closed off" to further suppletion, and may thus have yet more allomorphs governed by more peripheral heads. An adjective lacking a suppletive comparative (such as *big–bigger.* . .) can only be continued in the superlative by regular forms, while an adjective like *good–better.* . . could in principle maintain the comparative root (*best*) or be yet further suppletive (*optimus*; subject to the condition that further suppletion involves a portmanteau). Both patterns (ABB, ABC) are indeed attested, with further suppletion unsurprisingly the rarer of the two.

We have attempted, in this chapter, to express what we see as a tantalizing similarity across two phenomena that are otherwise completely unrelated, and we have proposed the principle in (1) to express this generalization. If we are right and the generalization is not spurious, then the observations here contribute to a characterization of a general property of language. Fundamentally, this property is the cyclic nature of derivations: rules apply in inner constituents before they apply in outer ones, with certain internal properties of inner/smaller constituents periodically "fixed" or "frozen" and thus inaccessible to later derivational steps (e.g., Bracket Erasure in Chomsky and Halle 1968, the PIC in Chomsky 2000, 2001, or the reformulated accessibility condition on spelled-out domains). Our proposal, in line with the references cited above, is that the formation of these impenetrable domains is suspended under certain conditions, but more specifically, that the conditions under which Domain Suspension applies themselves generalize across components, spanning the domains of morphophonology and syntax-semantics.

Notes

For feedback on some of the ideas presented here, we are grateful to audience members at *ETI 1* (McGill), ZAS (Berlin), *NELS 42* (Toronto), to seminar participants at UConn and MIT, and to the editors and reviewers for this book.

1. To keep within the scope (and page limits) of this chapter, we do not consider other domains here, including, for instance, interactions across phonological domains (as raised by reviewers).

2. Sources for these examples are: Persian–Mace 2003; Cimbrian–Schweizer 2008; Czech–Janda and Townsend 2000; Hungarian–Kiefer 2001; Latvian–University of Latvia 1999; Ubykh–Dumézil 1931.

3. That CMPR is cyclic does not follow from any prior assumptions, although it may be related to the assumption that category-changing morphology is cyclic (e.g., Embick 2010). Hints to that effect are provided by Russian comparatives, which are morphologically in essence adverbs (i.e., invariant short neuter adjectives) and the neutralization of the adjective/adverb distinction in English comparatives.

4. Note that (7a) is presented as a portmanteau (cumulative exponent), and thus represents Vocabulary Insertion at a node that dominates both the root and the comparative

head/feature. This can be achieved via insertion at nonterminal nodes (see Caha 2009; Radkevich 2010) or equivalently via the prior application of a rule of *Fusion* joining two terminal nodes into a single locus of exponence (Halle and Marantz 1993; Chung 2007). Theoretically, treating *opt-* as a portmanteau allows Vocabulary Insertion to target a node that is adjacent (both structurally and linearly) to SPRL, a condition on contextual allomorphy that is perhaps general (Embick 2010). Empirically, this explains why the *-iss-* element of the superlative (a reflex of the comparative) is missing in *opt-imus* (*opt-iss-imus), cf. *beatus–beat-ior–beat-iss-imus* = 'happy-CMPR-SPRL'. See Bobaljik 2012.

5. As we discuss below, the claim about the phasehood of a particular type of clausal complement is not about whether a structure involves raising, control, or ECM, but rather about the temporal and mood properties of the complement and how the complement combines with the matrix verb. To the extent these (in)dependencies correlate with properties such as raising, etc., we may use these familiar construction names to designate classes of clausal complements. But see note 8 for further details and qualifications.

6. Note that overt movement is subject to a general Last Resort condition that allows movement only when necessary for feature licensing. Since the subject of a raising infinitive requires Case licensing by the matrix T, movement is permitted under, for instance, the Last Resort condition in Bošković 2007, 610: X undergoes movement iff without the movement, the structure will crash (with crash evaluated locally). Similarly, successive cyclic *wh*-movement is allowed since it involves a feature licensing relation between an interrogative C and a *wh*-phrase.

7. Ora Matushansky (personal communication) suggests that the dynamic phasehood view, in particular the effect that the complement of a lexical head is a phase whatever its category/size, may be unified with the cycle-defining nature of category-changing morphology (see note 3). We leave pursuit of this intriguing idea for future work.

8. As mentioned in note 5, the claim that English raising infinitives are phases does not entail that raising constructions crosslinguistically are phases. Rather, phasehood is dependent on the selectional properties between a verb and its complement. In Wurmbrand, Alexiadou, and Anagnostopoulou 2012, we suggest, for instance, that raising constructions in Greek, Romanian, and Spanish are value-selected subjunctives or infinitives, and hence do not constitute phases, which, in contrast to English raising, allows phase-bound operations to apply across them. Furthermore, modal constructions, which can also involve raising (see Wurmbrand 1999; Bhatt 2000), show different QR properties than *seem* constructions in English, since modals originate in the tense domain and movement from the edge of the lower phase (Spec,*v*P) across a modal is allowed by Scope Economy.

12 Contextual Neutralization and the Elsewhere Principle

Karlos Arregi and Andrew Nevins

12.1 Introduction

Against the setting of this book, our aim is to contextualize the present chapter within the background of developments in phonological and morphological theory of the last forty-odd years, largely those arising from contributions by Morris Halle and his collaborators. The notions of specificity-based competition and blocking, with their indubitable Pāninian pedigree, found their way into modern generative linguistics with the introduction of the Elsewhere Principle in Kiparsky 1973, the goal of which was an attempt to reduce extrinsic ordering in Chomsky and Halle 1968. The intuition behind such a principle was that certain rules (or more broadly, operations that modify linguistic representations) will *always* take precedence over others, given their forms and the relationship of their forms to each other in terms of the fundamental notion of the subset relation from Set Theory. Kiparsky's contribution to phonological theory allowed researchers—and by hypothesis, language acquirers—to merely inspect the form of certain rules in order to determine their relative ordering.[1]

A similar paradigm shift (perhaps a term to be taken literally in this context) occurred within the view of Vocabulary Insertion as a procedural mapping from morphosyntactic terminals to phonological sequences. In Noyer 1992, vocabulary entries were rules with some extrinsic ordering among them, followed in turn by appeals to purely intrinsic ordering in DM, especially Halle and Marantz 1993, 1994, and Halle 1997a.[2]

The use of the Elsewhere Principle in DM is to regulate use of the *elsewhere item*, the least specified vocabulary entry that still does not constitute a superset or nonoverlapping set of the set of morphosyntactic features on the terminal node to be expressed. Consider for example the paradigm of third-singular pronouns in English. The masculine forms display a three-way case contrast (nominative *he*, accusative *him*, genitive *his*), and the feminine forms neutralize accusative and genitive forms (*her*), which are distinct from the nominative form (*she*). Thus, *her* is arguably an elsewhere feminine singular form, without any case specification. What prevents it from being used in the nominative contexts, then? The Elsewhere Principle, which dictates that the elsewhere form is only to be used when a more specific form is not found. In the case at hand, nominative *she* is more specific, blocking use of the elsewhere item *her*. The Elsewhere Principle is also known as the Subset Principle, specifically because it is implemented in terms of subset-based comparisons among sets of morphosyntactic feature-value pairs.

The notion of specificity is normally used in reference to the inherent morphosyntactic properties borne by the vocabulary entry itself. But at times, vocabulary entries can be *contextually restricted*, meaning that they can only be used in environments defined by adjacency or dominance in a local sense. As an example of this, consider the allomorphy between *destroy* and *destruct*, regulated by sisterhood to transitive v^*. In the examples that follow with the format $E \leftrightarrow M/X\_\_\_\_Y$, E is the exponent (the vocabulary entry), M is the morphosyntactic feature specification (MFS), and $X\_\_\_\_Y$ is the contextual restriction. We assume that both /dəstɹoj/ and /dəstɹʌkt/ are allomorphs of an abstract, categoryless root (in the sense of Arad 2003), denoted here as $\sqrt{\text{DESTRXYZ}}$:

(1) *Allomorphs of the root destroy/destruct, differing only in contextual restriction*
   a. /dəstɹoj/ $\leftrightarrow$ $\sqrt{\text{DESTRXYZ}}/v^*\_\_\_\_$
   b /dəstɹʌkt/ $\leftrightarrow$ $\sqrt{\text{DESTRXYZ}}$

The second item, /dəstɹʌkt/, is an elsewhere item—less specific with respect to context. As such, it occurs in all environments besides those with immediate sisterhood to v^*, including adjectives (*destructive, destructible*) nouns (*destruction*), and root compounds in which the root $\sqrt{\text{SELF}}$ blocks sisterhood with v^* in *self-destruct*.[3] The allomorph /dəstɹoj/ is only used in a limited/specialized environment.[4]

In the two examples we have examined thus far, one involved competition in terms of specificity of morphosyntactic features (*she* vs. *her*) while the other involves specificity in terms of context (*destroy* vs. *destruct*). As such there is no potential need to arbitrate between vocabulary entries in which one might be more specific for MFS while the other for context. But in fact precisely such a formulation already exists in the DM literature, one in which specificity in MFS takes precedence over specificity in context (Halle and Marantz 1993, 120–124):

(2) a *Underspecification*: The exponent in a vocabulary entry is eligible for insertion into a terminal node if the entry's MFS is a subset of the features in the terminal node, and if the contextual restriction of the former is compatible with the context of the latter.

    b. *Elsewhere Principle*: Where several entries meet Underspecification, the one matching the greatest number of features in the terminal node must be chosen.

    c. *Contextual Specificity*: Where several entries meet (2b), the one with the most specific contextual restriction must be chosen.

In other words, exponents whose MFS and/or context form a superset of the terminal node to be realized are immediately discarded (2a). Subsequently, the entry with the maximally matching MFS is chosen (2b), and only in the case of a tie is context appealed to (2c).

Interestingly, Halle and Marantz do not provide evidence for this prioritization. In fact, the phenomena discussed in the DM literature (including those in Halle and Marantz 1993) are typically consistent with this or the opposite preference of (2c) over (2b). In this chapter, we wish to reverse the importance (and indeed, computational priority) given to (2c) as opposed to (2b) above, thus placing Contextual Specificity *before* morphosyntactic specification, based on empirical arguments from Basque and Bulgarian, in which a morphosyntactic feature distinction ordinarily made—and indeed, one supported by ample vocabulary entries—is nonetheless jettisoned and neutralized in a particular environment.

The prediction of the "standard" prioritization, as schematized above, is that given Vocabulary Insertion alone (i.e., without the interference of impoverishment or other feature-modifying operations) defaults will never override specific entries, even if the default has a richer contextual restriction. The goal of this chapter is to provide evidence that Contextual Specificity takes precedence over the Elsewhere Principle, based on cases in which a featurally underspecified but contextually rich entry overrides more specific entries with a poorer contextual restriction. We provide two case studies as evidence: Basque pronominal clitics (section 12.2), and Bulgarian definite articles (sections 12.3 and 12.4). Before we turn to these case studies, we outline certain assumptions about the architecture of the grammar that will provide an important guide to our analysis of Basque and Bulgarian.

Within DM, it is generally agreed that impoverishment is early and linearization is late. The overall architecture of the postsyntactic morphological component in Arregi and Nevins 2012, which we adopt here, is depicted in figure 12.1. The main points that run through our architecture that are

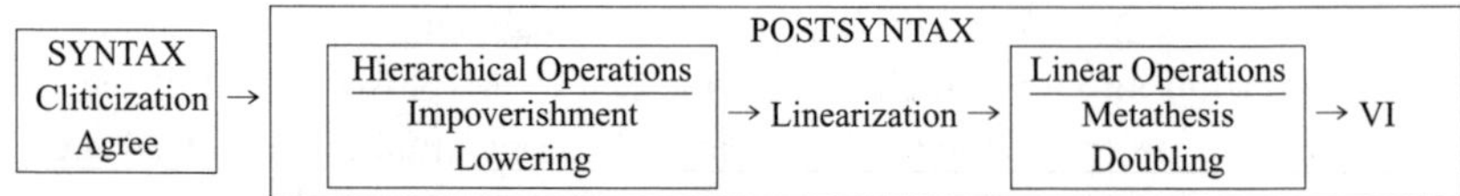

Figure 12.1
The structure of Spellout, as proposed by Arregi and Nevins 2012

important for this chapter are the hypothesis that impoverishment is, as often as possible, principled, while vocabulary entries—later down in the chain—are more idiosyncratic and language particular. For this reason, highly specific contextual effects on allomorph selection are likely to be the provenance of Vocabulary Insertion, and thus more likely to change, cross-dialectally, than impoverishment-based ones. In addition, operations that apply before linearization, such as impoverishment and lowering, are defined in terms of hierarchical relations, while postlinearization processes such as metathesis (section 12.2) are sensitive to linear order.

12.2 Contextual Neutralization in Basque Pronominal Clitics

In Biscayan Basque, pronominal enclitics exhibit case contrasts that are neutralized in proclitic position. This, we argue, is a case of contextual neutralization in the sense defined above: case-neutral vocabulary entries specific to proclitic contexts override case-specific entries. This provides the first piece of evidence that Contextual Specificity takes precedence over the Elsewhere Principle in determining competition at Vocabulary Insertion. For reasons of space, we omit many of the details of the analysis, and the reader is referred to Arregi and Nevins 2012 for extensive argumentation and comparison with alternative accounts, as well as further illustration of the phenomena discussed here.

Since most verbs in this language lack finite forms, finite clauses are typically headed by an auxiliary that cross-references phi-features of absolute, dative, and ergative arguments in the clause. The following example from the Biscayan variety of Ondarru is illustrative:[5]

(3) Neu-k seu-∅   ikus-i *s*   *-atxu*   *-t.*
    I.ERG you(SG).ABS see.PFV **CL.ABS.2SG -PRS.2SG -CL.ERG.1SG**
    'I've seen you(SG).'   Ondarru

In this auxiliary, the root *-atxu-* encodes present tense, in addition to agreement (second singular) with the absolutive argument. Following Arregi and Nevins

2012, we take this exponent to be the realization of T. Pronominal clitics are illustrated by the second singular absolute proclitic *s-* and the first singular ergative enclitic *-t*. The following descriptive template displays the relative position of the different morphemes in the auxiliary:

(4) *Morphemes in the Basque auxiliary*[6]
    abs proclitic – T (root) – dat enclitic – erg enclitic – complementizer

We assume that this is little more than a descriptive template of surface auxiliary form, derivable in a principled way from the syntax of the morphemes involved. Since these details are not directly relevant to the issues of interest here, we abstract away from them in this chapter.

As shown in (4), absolutive clitics precede the root, while dative and ergative clitics follow it (with important exceptions discussed below). The following examples from Ondarru provide further illustration of first singular pronominal clitics in all three cases:

(5) *Dative enclitic*

    Neu-ri   emo-n   *d*   *-o*   **-sta**   -∅   (>emo-sta).
    me-DAT give-PFV CL.EP -PRS.3SG **-CL.DAT.1SG** -CL.ERG.3SG
    'She's given it to me.'   Ondarru

(6) *Ergative enclitic*

    Neu-k emo-n   *d*   *-o*   *-tza*   **-t**   (>emo-tzat).
    I-ERG give-PFV CL.EP -PR.3SG -CL.DAT.3SG **-CL.ERG.1SG**
    'I've given it to him.'   Ondarru

(7) *Absolutive proclitic*

    Neu-∅ etorr-i   *n*   *-as*.
    I-ABS come-PFV **CL.ABS.1SG** -PRS.1SG
    'I've come.'   Ondarru

These examples illustrate the fact that enclitics display case contrasts: the first singular dative enclitic is *-sta* (5), while its ergative counterpart is *-t* (6). Table 12.1 provides a full paradigm of the surface form of pronominal clitics in the Biscayan dialectal area (this figure abstracts away from allomorphy and dialectal variation not directly relevant here).[7] As shown in this paradigm, enclitics in all phi-feature combinations contrast in case. Despite significant variation in surface form, these case contrasts are present throughout the Biscayan dialect (de Yrizar 1992). Absolutive proclitics are illustrated in (7);

Table 12.1
Pronominal clitic forms in Biscayan Basque

	Proclitics		Dative enclitics		Ergative enclitics	
	Singular	Plural	Singular	Plural	Singular	Plural
First	n-	g-	-sta	-sku	-t	-gu
Second	s-	s-. . .-e	-tzu	-tzue	-su	-sue
Third	—	—	-tza	-tze	-Ø	-e

their exponents are given a case-neutral label in table 12.1, since, as argued below, proclitics of the same form can also double ergative and dative arguments.

Contextual neutralization of case can be observed in certain forms where the proclitic doubles an ergative or dative argument (instead of the expected absolutive). This is due to the application of different processes that displace or copy an enclitic into auxiliary-initial position:

(8) *Ergative Metathesis: Ergative in proclitic position*

Neu-k emo-n   ***n***   *-e*   *-tza*   *-n*   (>emo-netzan).
I-ERG give-PFV **CL.ERG.1SG** -PST.3SG -CL.DAT.3SG -CPST
'I gave it to her.'   Ondarru

(9) *Ergative Doubling: Ergative in enclitic and proclitic position*

s   *-eu*   *-sku*   **-su**   *-n*
CL.ERG.2SG -PST.3SG -CL.DAT.1PL **-CL.ERG.2SG** -CPST
Alboniga (de Yrizar 1992, vol. 1, 470)

(10) *Dative Doubling: Dative in enclitic and proclitic position*

Ar-ek   ne-ri   sagarr-a   emu-n
he-ERG me-DAT apple-ABS.SG give-PFV

n   *-o*   **-sta**   **-Ø**   *-n.*
CL.DAT.1SG -PST.3SG **-CL.DAT.1SG** -CL.ERG.3SG -CPST
'He gave me the apple.'   Oñati (Rezac 2008a, 710)

In addition to causing a change in position, these processes also have an effect on the form of the clitic. All proclitics in these examples, regardless of case, are realized as shown in the case-neutral proclitic paradigm in table 12.1. This is perhaps most clearly seen in the doubling examples (9) and (10), where the enclitic copy has the expected case-specific form (see table 12.1), but the auxiliary-initial copy has the case-neutral proclitic form. Thus, the Basque clitic paradigm displays contextual neutralization: the case contrasts visible in enclitic position in (5)–(6) are neutralized in proclitic position in (8)–(10).

Note, furthermore, that these are bona fide ergative and dative proclitics, not analyzable in terms of absolutive case assignment to an external or Goal argument, as evinced by the fact that the strong pronouns cross-referenced by the proclitics in (8) and (10) have ergative and dative case, respectively. In addition, in the doubling cases in (9) and (10), the auxiliaries themselves contain another clitic exponent with unmistakably ergative or dative form cross-referencing the same argument. Another indication that the proclitic in these examples is not absolutive is that, if that were the case, we would expect it to trigger agreement in T, but, as shown by the glosses, the form of T in these auxiliaries is specific to third-person agreement, not first- or second-person.[8]

Ergative Metathesis, also known as *ergative displacement* in the literature (Heath 1976; Bonet 1991; Laka 1993; Albizu and Eguren 2000; Fernández and Albizu 2000; Rezac 2003), occurs in all dialects of Basque, and is limited to contexts where tense is nonpast and the absolutive argument is either third singular or altogether absent. Following Laka 1993, we propose in Arregi and Nevins 2012 a postsyntactic displacement analysis in which the ergative cliticizes to enclitic position in the syntax, but is displaced to auxiliary-initial position prior to Vocabulary Insertion. This postsyntactic displacement is triggered by Noninitiality, a constraint on the linearization of T that prevents it from surfacing in initial position in the auxiliary. In auxiliaries with absolutive clitics (e.g., (7)), the latter are linearized to the left of T, and therefore no postsyntactic repair operation is needed to satisfy Noninitiality. In the absence of an absolutive clitic, certain repair operations apply to shield T from the left edge. Ergative Metathesis, which applies under the conditions specified above, is one of those operations: by displacing the ergative clitic to the left of T, the structure satisfies Noninitiality. In other contexts (e.g., in the present tense, or in the absence of an ergative clitic), an epenthetic morpheme is inserted to satisfy the constraint. This epenthetic morpheme is exemplified in (5)–(6) (glossed as "CL.EP").

In Arregi and Nevins 2012, chap. 5, we implement this displacement operation in terms of Harris and Halle's (2005) Generalized Reduplication formalism. This implementation allows us to extend the analysis to Ergative and Dative Doubling. These processes, which have a more restricted dialectal distribution (Fernández 2001; Fernández and Ezeizabarrena 2003; Rezac 2008a, 2008b),[9] apply instead of Ergative Metathesis in a subset of the contexts where the latter is expected to apply, and are similarly triggered by the need to shield T from the left edge of the auxiliary.

The generalizations about the form of (Biscayan) Basque pronominal clitics uncovered above can be implemented in terms of vocabulary entries for proclitics that are case neutral (hence resulting in contextual neutralization of case) competing with entries that are specific for dative and ergative case. *In other words, there are no clitics specified as absolutive in Basque.* In the case of first-person singular, the following entries have these properties (entries for other phi-features are similar):[10]

(11) *Vocabulary entries for first singular clitics in Biscayan Basque*
    a. /n/ ↔ [first, singular]/______T
    b. /t/ ↔ [first, singular, ergative]
    c. /sta/ ↔ [first, singular, dative]

The entry for *n-* in (11a) is specific to proclitic position (before T), and is therefore not relevant in the realization of clitics following T.[11] Thus, enclitics are realized as ergative *-t* (11b) or dative *-sta* (11c)—that is, the realization of enclitics results in the observed case contrast in this position. In proclitic position, on the other hand, both the case-neutral proclitic entry (11a) and one of the two case-specific entries are candidates for insertion (as long as the proclitic is ergative (11b) or dative (11c); if absolutive, only (11a) is eligible). Given our hypothesis that Contextual Specificity takes precedence over the Elsewhere Principle, the correct prediction is that case-neutral (11a) is inserted, since, despite the fact that its MFS is a subset of the MFS of the case-specific entries, its contextual restriction (before T) is richer than the null contextual restriction in the other entries.[12]

Before we turn to further evidence from Bulgarian, we need to rule out other possible analyses of contextual neutralization in Basque pronominal clitics. Relying on certain phonological similarities between dative and ergative enclitics evident in table 12.1, one might argue that they are derived by phonological processes from common case-neutral underlying enclitic forms. Under this analysis, the paradigm would not display contextual neutralization, since no case contrasts would be posited in enclitic position in the first place. Although these phonological similarities provide evidence for a common diachronic analysis of the form of dative and ergative enclitics, we argue in Arregi and Nevins 2012, 127–132, that they do not justify such an analysis in the synchronic grammar of Biscayan Basque. These similarities are greater in other dialects—for instance, first- and second-person enclitics have case-neutral forms in the standard dialect (Batua; see Hualde 2003), and in several spoken varieties of other dialects. On the other hand, enclitics in many other varieties do have case-based contrasts not attributable to synchronic phonology. For instance, in Souletin, the first plural ergative enclitic is *-gü*, while its dative counterpart is *-kü* (de Yrizar 2002), but their phonological contexts (e.g., *dü-gu, zai-kü*) do not warrant an analysis in terms of (de)voicing. A similar argument for contextual neutralization can also be made based on these latter dialects, but we concentrate on Biscayan here because these enclitic case contrasts are clearer in this dialect.

A different type of alternative analysis would, like ours, account for the facts in terms of contextual neutralization, but would rely on independently motivated mechanisms instead of a change in the way that Vocabulary Insertion determines competition. First, one could add some restriction to the case-specific entries that excludes them from auxiliary-initial position. Since the case-specific exponents in (11b) and (11c) only appear in enclitic position, it is tempting to add the contextual restriction T____ to them, thus making them ineligible for insertion in proclitic position. This would make competition for the realization of proclitics a nonissue and the proposed change to Vocabulary Insertion unnecessary. Unfortunately, given certain well-founded assumptions in current work on Distributed Morphology, this alternative analysis does not make the correct predictions. Under the hypothesis that a contextual restriction can only make reference to features on adjacent terminal nodes (Embick 2010), it would wrongly predict that both case-specific entries in (11) are restricted to clitics that are right-adjacent to T. Although ergative clitics can be right-adjacent to T (in the absence of a dative clitic), they need not be, as shown in (6), where the ergative clitic follows a dative clitic. However, the form of the ergative clitic is identical in both cases and clearly not dependent on how close it is linearly to T.

A second alternative analysis along similar lines involves impoverishment. This type of rule, which either deletes a feature or changes it to an unmarked value, is often used in the DM literature in order to account for contextual neutralization facts. For instance, several authors propose an impoverishment-based analysis of spurious *se* in Spanish (Bonet 1991, 153–173; Halle and Marantz 1994; Nevins 2007, 274–283). In this language, the dative clitic is *le* (*les* in the plural), except in the context of an accusative clitic, in which case it is realized as *se*, which is syncretic with the reflexive/impersonal pronoun. Thus, a contrast between dative and reflexive clitics is neutralized in the context of accusative clitics. Under the assumption that the *se* exponent lacks some feature that dative *le* is specified for (person in Bonet 1991 and Nevins 2007; case in Halle and Marantz 1993), we can account for this case of contextual neutralization by impoverishing that feature in a dative clitic if it occurs in the same cluster as an accusative clitic.

This would suggest an alternative account of the contextual neutralization facts in Basque pronominal clitics in which an impoverishment rule deletes case features (or changes them to unmarked absolutive) in proclitic position. Since ergative and dative clitics surface in proclitic position due to the application of

Ergative Metathesis/Doubling and Dative Doubling, this impoverishment rule would have to apply after these operations affect the position of these clitics. Given the restrictive and modular architecture of the postsyntactic component proposed in Arregi and Nevins 2012 and briefly reviewed in section 12.1, this is not a viable option for the Basque facts. In particular, impoverishment rules, which are typically not sensitive to morpheme order, apply before linearization, and rules that alter the linear order of morphemes, such as the metathesis and doubling rules discussed above, apply after linearization. This predicts that impoverishment rules systematically apply prior to metathesis and doubling processes, and therefore have the potential to feed or bleed them, a prediction that we provide evidence for in Arregi and Nevins 2012, chap. 6.

To summarize so far, contextual neutralization facts in Basque pronominal clitics provide evidence for our hypothesis that Contextual Specificity takes precedence over the Elsewhere Principle at Vocabulary Insertion. In the following section, we present additional evidence from the realization of the definite article in Bulgarian.

12.3 Contextual Neutralization in Bulgarian Definite Articles

The Bulgarian definite article paradigm displays phi-feature-based contrasts that are neutralized in certain phonologically defined environments. As with Basque pronominal clitics, we propose an analysis in which this case of contextual neutralization is the result of underspecified vocabulary entries with (phonological) contextual restrictions overriding entries that are phi-feature specific but lack a contextual restriction.

In Bulgarian, the definite article surfaces as an enclitic attaching to either the head noun or the first noun modifier in the DP, whichever comes first. The following are relevant examples from Embick and Noyer 2001, 568–569:[13]

(12) a. kniga-**ta**
          book.FEM.SG-DEF
          'the book'
      b. xubava-**ta** kniga
          nice-DEF   book.FEM.SG
          'the nice book'
      c. mnogo starij-**a** teatər
          very   old-DEF theater.MASC.SG
          'the very old theater'

A lot of the literature on the Bulgarian definite article concentrates on accounting for its position within the structure of DP (see, among others, Franks 2001;

Embick and Noyer 2001; Dost and Gribanova 2006).[14] We adopt Embick and Noyer's (2001) analysis, according to which the article is generated in the syntax as the head of DP, and its surface position is due to postsyntactic lowering:

(13) *Structure of (12a–b) after lowering*

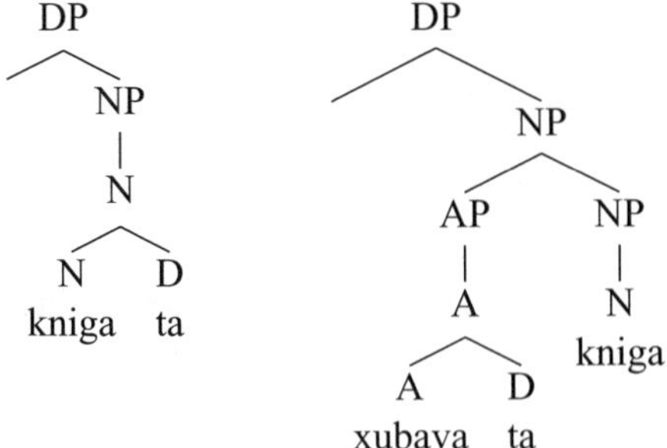

As illustrated in these examples, lowering, like head movement, results in X^0-adjunction. In terms of the architecture of the grammar in figure 12.1, lowering is a hierarchically defined postsyntactic operation that applies before linearization and is thus distinct from other morpheme-displacement operations such as syntactic head movement and postlinearization metathesis.[15]

The form of the article, which is the main topic of this section, is in part dependent on the gender and number features of the word it attaches to. For instance, the adjective *starij-* in (12c) is masculine singular, due to agreement with the noun *teatər*, and, accordingly, the form of the definite article *-a* attached to the adjective is specific to masculine singular hosts. We assume that the source of this sensitivity to (masculine/feminine/neuter) gender and (singular/plural) number features in the article are due to DP-internal agreement—that is, at the point of Vocabulary Insertion, the article (in addition to its host) is specified for valued phi-features. We provide an analysis of DP-internal agreement in Bulgarian in section 12.4.

Both morphosyntactic and phonological factors determine the form of the definite article. The following are the relevant generalizations (Harizanov and Gribanova 2011):[16]

(14) *The realization of the definite article in Bulgarian*
   a. If the host ends in the vowel *a* or *o*, then the definite article is realized as *-ta* or *-to*, respectively; otherwise,
   b. if the definite article is singular masculine, then it is realized as *-a*,
   c. if the definite article is singular feminine, then it is realized as *-tá*,
   d. if the definite article is singular neuter, then it is realized as *-to*, and
   e. if the definite article is plural, then it is realized as *-te*.

While (14b–e) make reference to the phi-features of the article, (14a) makes reference to the phonological features of its context. As stated clearly in (14), the latter takes precedence over the former, hence this is a case of contextual neutralization: phi-featural distinctions in the article that are otherwise visible in the paradigm are neutralized in the specific context of a host that ends in the vowel *a* or *o*.

The following examples illustrate the generalizations above.[17] First, masculine singular nouns that end in a consonant take the *-a* allomorph of the article:

(15) *Masculine singular nouns ending in C: -a*
    məʒ-**a** 'the man'   sto**l-a** 'the chair'

Masculine singular nouns can also end in the vowels *a* and *o*, in which case they take the *-ta* and *-to* allomorphs of the article, respectively:

(16) *Masculine singular nouns ending in a: -ta*
    baʃta-**ta** 'the father'   sədija-**ta** 'the judge'

(17) *Masculine singular nouns ending in o: -to*
    tatk**o-to** 'the father'   djad**o-to** 'the grandfather'

Feminine singular nouns typically end in the vowel *a* and accordingly take the *-ta* allomorph of the definite article:

(18) *Feminine singular nouns ending in a: -ta*
    ʒena-**ta** 'the woman'   staja-**ta** 'the room'

However, a few feminine singular nouns end in a consonant. These take a different allomorph of the article, *-tá*, specific to feminine singular and distinguishable from *-ta* in that it attracts stress:

(19) *Feminine singular nouns ending in a consonant: -tá*
    mlados**t-tá** 'the youth'   doblest-**tá** 'the valor'   cev-**tá** 'the barrel'

Neuter singular nouns always end in a vowel, and regardless of the features of this vowel, the article is realized as *-to*:

(20) *Neuter singular nouns: -to*
    sel**o-to** 'the village'   det**e-to** 'the child'
    menj**u-to** 'the menu'   taks**i-to** 'the taxi'

Bulgarian plural nouns are formed on the basis of several allomorphs of a nominal plural suffix.[18] Nouns with the plural suffixes ending in *a* take the *-ta* allomorph of the definite article, as expected:

(21) *Plural nouns ending in a: -ta*
    brat-j**a-ta** 'the brothers'   krai-ʃt**a-ta** 'the ends'

Table 12.2
Contextual neutralization in Bulgarian definite articles

	Masculine	Feminine	Neuter	Plural
Host ends in *a*	baʃta-**ta**	ʒena-**ta**		brat-ja-**ta**
Host ends in *o*	tatk**o-to**		sel**o-to**	
Other hosts	məʒ-**a**	mladost-**tá**	dete-**to**	narod-i-**te**

With plural suffixes that end in a vowel other than *a*, the allomorph of the definite article is *-te*:

(22) *Plural nouns ending in a vowel other than a: -te*
    narod-**i-te** 'the peoples'   məʒ-**e-te** 'the men'

These generalizations are succinctly summarized in table 12.2, which clearly represents the fact that the paradigm exhibits contextual neutralization. The bottom row exemplifies the four-way gender/number contrast that the form of the definite article is sensitive to. The other rows show that these phi-featural contrasts are neutralized in the context of hosts ending in the vowels *a* or *o*.

Contextual neutralization in Bulgarian definite articles can thus be accounted for in a way parallel to Basque pronominal clitics. First, we propose the following vocabulary entries (adapted from Harizanov and Gribanova 2011):[19]

(23) *Vocabulary Entries for Bulgarian definite articles*
    a. /ta/ ↔ [definite] / [/-a/] _____
    b. /to/ ↔ [definite] / [/-o/] _____
    c. /a/ ↔ [definite, singular, masculine]
    d. /tá/ ↔ [definite, singular, feminine]
    e. /to/ ↔ [definite, singular, neuter]
    f. /te/ ↔ [definite, plural]

With hosts other than those ending in *a* or *o*, only the phi-specific exponents in (23c–f) are eligible for insertion. As a consequence, in these contexts, the phi-featural contrasts are visible. Compare, for instance, masculine singular *məʒ-a* (15) and feminine singular *mladost-tá* (19):

(24) [N [D definite, singular, masculine]]   [N [D definite, singular, feminine]]
        məʒ   **-a**   mladost   **-tá**

On the other hand, in the context of a host ending in *a* or *o*, these phi-specific entries compete with the phi-featurally underspecified but contextually rich entries for *-ta* (23a) and *-to* (23b). Given that Contextual Specificity takes precedence over the Elsewhere Principle, the latter entries are chosen for insertion, as illustrated here with masculine singular *baʃta-ta* (16) and feminine singular *ʒena-ta* (18):

(25) [N [D definite, singular, masculine]]   [N [D definite, singular, feminine]]
      baʃta   **-ta**   **ʒena**   **-ta**

The result is contextual neutralization: the phi-featural contrasts illustrated in (24) are neutralized in the context exemplified in (25).

The Bulgarian allomorphy data are representative of a larger class of phenomena in which phonologically sensitive considerations seem to trump morphosyntactic specificity. In this light, they are reminiscent of definite article allomorphy in well-known cases such as Spanish and French, in which vowel-initial nouns may take articles of the "wrong" gender. To take the simplest case among these, nouns beginning with stressed *á* in Spanish have the gender of their definite article *neutralized* to the form usually reserved for masculines, namely *el*. This constitutes a case of contextual neutralization in which the masculine/feminine distinction in the definite article, otherwise robust and based purely on morphosyntactic features, is jettisoned in favor of a context-sensitive form that looks at the phonology of the stem (see Nevins 2011c for an overview).

In this light it is interesting to compare our proposed revision of the Elsewhere Principle of Vocabulary Insertion to the novel two-step approach to Vocabulary Insertion proposed in Svenonius 2012b, on the basis of definite article in French, which is also phonologically sensitive. Svenonius's proposal shares with ours the fact that the first step of Vocabulary Insertion is purely to eliminate superset candidates of vocabulary entries whose specification includes features not in the terminal node being matched. It also shares with ours the hypothesis that phonological factors may be referred to—and decisive in allomorph selection—before maximal subset considerations are taken into account. One of the differences, however, is that Svenonius's proposal attempts to cleanly partition the two steps of Vocabulary Insertion into superset elimination (called 'L-Match') and a phonologically optimizing stage. It is on this latter point that we diverge from Svenonius, pointing specifically to the Bulgarian case at hand. Consider the feminine allomorphs, either stressed *-tá* or unstressed *-ta*, the latter chosen when the stem ends with the vowel *a*. It is not clear what types of phonotactic or metrical pressures would force the preference for *-ta* over *-tá* following an *a*. Coupled with the fact that the Basque case discussed above does not involve phonological sensitivity, we contend that the correct characterization of Vocabulary Insertion is indeed one in which context sensitivity trumps morphosyntactic specification, but where context sensitivity need not be limited to purely phonologically optimizing considerations. Nonetheless, we contend that Svenonius's division of labor of Vocabulary Insertion into two separate stages, in which a principle of Phonology-Free Syntax is upheld in the sense that vocabulary entries themselves never directly mention phonological context, rather leaving such choices up to the grammar, as it were, constitutes an interesting move in the direction of modularization in the general spirit of DM and is worthy of extensive further comparison with the proposal we have developed here.

12.4   The Syntax and Postsyntax of Agreement in Bulgarian Definite Articles

A crucial assumption in the analysis of Bulgarian given above is that the definite article is specified for phi-features as a consequence of DP-internal agreement. This assumption is challenged by Harizanov and Gribanova (2012), who claim that the phi-featural factors that (partially) determine the form of the definite article are not due to features in the article itself, but to features in its host. For instance, under this analysis, the masculine singular *məʒ* in (24) selects the masculine singular allomorph *-a* of the definite article not because the article itself is specified for these features, but because the host *məʒ* is. Thus, in Harizanov and Gribanova 2012, the phi-features in the entries for (23c)–(23f) are not part of the MFS, but part of the contextual restriction. Under this view, the Bulgarian definite article paradigm does not constitute a case of contextual neutralization, since all the allomorphs have identical MFS ([definite]) and only differ in their contextual restriction.

Harizanov and Gribanova's (2012) argument is based on DPs with coordinated adjectives:

(26) bălgarskij-**a**   i   ruski   narod-i
     Bulgarian.MASC.SG-DEF and Russian.MASC.SG nation.MASC-PL
     'the Bulgarian and Russian nations'
     (= the Bulgarian nation and the Russian nation)   (Harizanov and
                                                               Gribanova 2012, 9)

What is interesting about this type of example is that the coordinated singular adjectives do not agree in number with the plural noun. Since, as indicated by the meaning, the syntactic scope of the definite article is the entire DP, we might expect the definite article to agree with the plural noun. This is not the case: the article in this example, which is attached to the adjective in the first conjunct (i.e., the first noun modifier in the DP), is realized by the *singular* masculine allomorph *-a*, not plural *-te*. One might be tempted to conclude that this is due to some sort of closest conjunct agreement with the leftmost adjective. This does not seem to be the case, since an adjective with a similar syntactic scope as the article in (26) does agree with the noun:

(27) prijatelsk-**i**-te   bǎlgarski   i   ruski   narod-i
     friendly-PL-DEF Bulgarian.MASC.SG and Russian.MASC.SG nation.MASC-PL
     'the friendly Bulgarian and Russian nations'   (Harizanov and
                                        Gribanova 2012, 10)

Taking (27) as representative of the agreement properties of items that have scope over the entire DP, Harizanov and Gribanova (2012) conclude that the singular allomorph -*a* of the article in (26) cannot be due to agreement (i.e., the article is not specified for phi-features); rather, it is due to contextual allomorphy conditioned by the singular adjective it is attached to.

We do not think that this conclusion is warranted, since it rests on the assumption that the agreement properties of agreeing items must be completely determined by their syntactic position and that therefore postsyntactic processes cannot have an effect on agreement. Recent literature on the topic suggests that this is not the case, and that postsyntactic properties of structures do indeed have an effect on agreement. Specifically, both Arregi and Nevins (2012, 81–88) and Bhatt and Walkow (forthcoming) argue (on quite different grounds) that agreement proceeds in two steps: agreement is established in the syntax, but implemented in the postsyntactic component, with the potential to be affected by information only available at this point in the derivation. We propose that the agreement asymmetry observed in (26) and (27) is due to this two-step procedure. The main motivation in the works cited above for splitting agreement into a syntactic step and a postsyntactic one is to account for phenomena that bear the structural signature of syntactic Agree, yet actual feature valuation is affected by postsyntactic operations (impoverishment in Arregi and Nevins 2012, and linearization in Bhatt and Walkow, forthcoming). This is, we claim, what accounts for the differing behavior of the article in (26) and the first adjective in (27): although they are in parallel structural configurations relevant for agreement in the syntax, the article, but not the adjective, is subject to postsyntactic displacement that alters this configuration and thus has an effect in the postsyntactic implementation of agreement. In particular, attachment of the article to the adjective in the first conjunct in (26) alters the locality relations with its potential agreement controllers in such a way that its feature values are copied from its (postsyntactic) sister adjective instead of the noun.

Before we spell out the details of our analysis, we need to make explicit our assumptions about the syntax of coordination, which is in part responsible for the mismatch in number between the coordinated singular adjectives and the plural head noun observed in (26)–(27). First, we assume an asymmetric analysis of coordination where a coordinating particle heads a phrase that we label "&P"; the coordinated elements fill the specifier and complement positions of this head (Munn 1992, Johannessen 1993, and much subsequent work). The conjoined adjectives in (26)–(27) thus have the following structure in the syntax:

(28)

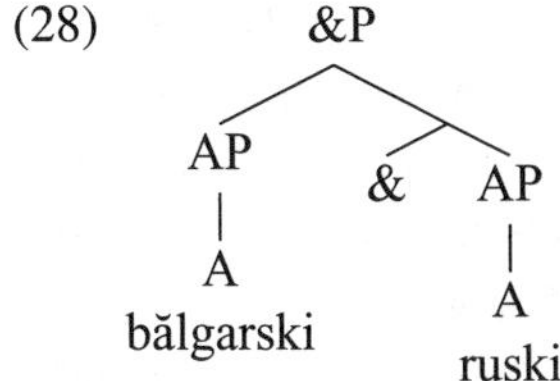

This entire phrase occupies the same position as AP modifiers—that is, it is adjoined to NP (see below for detailed structures). Second, &P undergoes DP-internal agreement with the head noun, in the manner specified below for DP-internal modifiers. Thus, &P in (26)–(27) is specified as masculine plural. On the other hand, the phi-feature values of each conjoined adjective are determined by agreement resolution within coordination (Corbett 1983), which in these examples results in masculine singular.[20] This accounts for the number mismatch mentioned above. Finally, lowering, as a postsyntactic operation, is not subject to Ross's (1967) Coordinate Structure Constraint, which accounts for the attachment of the article onto the first conjoined adjective in (26).

Our specific implementation of the two-step procedure for agreement is as follows.[21] An agreement controller is a *probe* (in the sense of Chomsky 2000) with unvalued phi-features. A probe establishes an *Agree-Link* relation with suitable *goals* (elements with matching features that might be valued or not). Crucially, Agree-Link is an abstract relation between two nodes, and does not effect feature valuation (this is accomplished by postsyntactic Agree-Copy, as discussed below). In DP-internal agreement, a probe establishes Agree-Link with all phi-feature-bearing elements in its c-command domain (this is Multiple Agree, in Hiraiwa's (2001) sense; see also van Koppen 2005 for Multiple Agree in coordinate structures). In the Bulgarian examples above, the article in (26) and first adjective in (27) establish the following Agree-Link relations (denoted by arrows):[22]

(29) *Agree-Link relations in (26) and (27)*

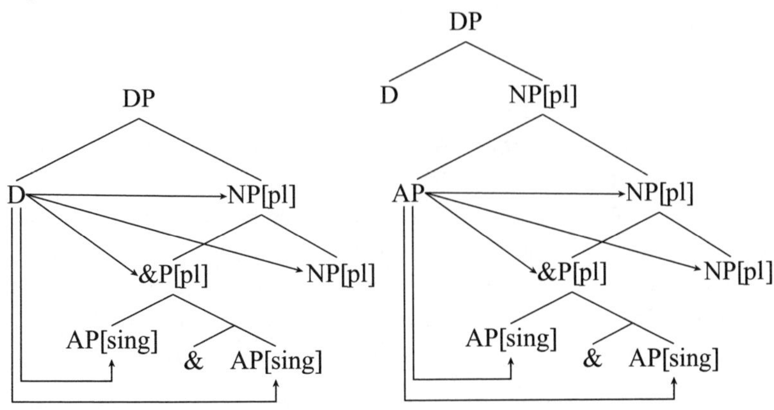

In both examples, the probe is agree-linked with plural goals (NP and &P) as well as singular goals (the AP conjuncts). Feature valuation is accomplished by *Agree-Copy* in the postsyntactic component, which copies feature values from the goal to the c-commanding probe. In cases with more than one goal agree-linked to a probe, features from the closest goal are copied, where *closest* is defined by standard locality conditions stated in terms of c-command and dominance (Fitzpatrick 2002 and references cited there): given a node x that c-commands nodes y and z, y is closer to x than z iff y c-commands or dominates z. In (27) (see rightmost tree in (29)), the closest goal to the topmost AP probe is its sister NP, which results in plural agreement. In effect, this analysis imposes standard locality conditions on Agree in the postsyntactic component, rather than the syntax: the probe is agree-linked with several goals in the syntax, and the most local one is selected postsyntactically by Agree-Copy.

This separation of agreement into syntactic Agree-Link and postsyntactic Agree-Copy correctly predicts singular agreement on the article attached to the first conjunct in (26). As shown in (29), the D probe in (26) is agree-linked with the same goals as the topmost adjective in (27). However, in the postsyntactic component, lowering alters the structure by attaching D to the leftmost conjunct.

(30) *Structure of (26) after lowering*

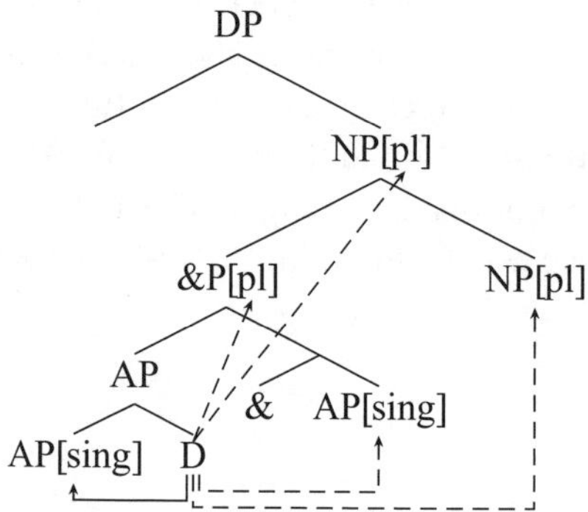

This in effect undoes the Agree-Link relations between the D probe and all goals except for the leftmost AP conjunct, since the latter is the only goal c-commanded by D (this is indicated in (30) by dashed vs. solid lines). As a consequence, Agree-Copy, which applies after lowering, copies the feature values from this AP, and D surfaces with singular number.

Thus, the crucial difference between the leftmost adjective in (27) and the article in (26) is that the structural position of the latter is changed in the postsyntactic component, which alters the locality relations relevant to Agree-Copy. Therefore, we contend, the phi-featural asymmetry observed in (26)–(27) is not a sign that the Bulgarian definite article does not agree; rather, it is a consequence of differences in the postsyntactic derivation of articles and adjectives that have an effect in the way that agreement is implemented in the postsyntactic component.

As in Arregi and Nevins 2012 and Bhatt and Walkow, forthcoming, the analysis detailed above is based on the idea that feature valuation is (at least in some cases) postsyntactic. However, there are important differences between the three analyses. One of these differences has to do with the location of Agree-Copy in the postsyntactic derivation. In our analysis of DP-internal agreement in Bulgarian, Agree-Copy applies after lowering, and the fact that the former is defined in purely hierarchical terms of c-command and dominance suggests that it precedes linearization, as proposed in Arregi and Nevins 2012. On the other hand, feature valuation in Bhatt and Walkow, forthcoming, is sensitive to linear order and therefore must apply after linearization. Whether the three analyses can be put together into a more comprehensive theory of agreement phenomena is a question that we leave for future research, but we note here that these differences in analysis are to a great extent due to the fact that they are designed to account for different phenomena: multiple agreement with absolutive and dative arguments in Basque in Arregi and Nevins 2012, closest conjunct agreement in Hindi in Bhatt and Walkow, forthcoming, and agreement fed by postsyntactic lowering in Bulgarian in the present chapter. It thus might well be the case that feature valuation is not a single operation and is in fact distributed in different stages of the postsyntactic derivation, with the concomitant expectation that crosslinguistic differences might be due to variation in the application of this postsyntactic process. In this light, it is worth noting that even within a uniform empirical domain, namely, closest conjunct agreement, the recent literature (van Koppen 2005; Bhatt and Walkow, forthcoming; Marušič, Nevins, and Badecker, forthcoming) reveals variation across languages (or even within a single language) in the factors that determine feature valuation (c-command, dominance, linear order, or even feature specificity in vocabulary entries).

12.5 Conclusion

We have shown that the Basque clitic system, with a three-way case distinction available among its clitics, nonetheless abandons the full utilization of such distinctions in a specific morphosyntactic context, namely left-adjacency to the auxiliary root. The fact that ending up in this highly "contingent" environment nonetheless trumps the otherwise applicable decision based on closest matching feature specifications constitutes an argument that neutralization may occur during VI as a consequence of how exactly specificity is evaluated. The necessity of such context sensitivity as an early step in eliminating candidates for VI can then be extended to Bulgarian, in which the four-way gender-number distinction among allomorphs of the definite article is nonetheless jettisoned given a categorical context variable with a specific phonological shape. This analysis of sensitivity of allomorph selection to phonological form—while perhaps seemingly "Talmudic" in terms of its overall point about the details of a very specific DM-internal mechanism—is based on potentially important empirical observations that only arise within the context of a theory, and we wish to reflect on how incremental advances of this sort are necessary to constantly reevaluate how Vocabulary Insertion—arguably the singly most irreducible property of DM—works in its gory details. As vocabulary entries are one of the most variable aspects of human language, one does not always come across cases that decidedly favor one formulation over another. Nonetheless, a focus on the specific properties of how disjunctive ordering is determined is among the many steps necessary as part of the overall broad research program pioneered by Morris Halle and Alec Marantz, the goal of which is to examine the organization of the morphological component in terms of computations on representations enacted by mechanisms that are distributed and shared across various modules of the grammar.

Notes

*Thanks to Boris Harizanov, Vera Gribanova, Peter Svenonius, and other participants at the Stanford Workshop on Locality and Directionality at the Morphosyntax-Phonology Interface, and to the members of GELA-Rio for excellent discussion. Special thanks to the editors, Ora Matushansky and Alec Marantz, for their encouragement and for organizing this endeavor.

1. One of Kiparsky's subsequent breakthroughs, along similar lines, is found in Kiparsky 1982, in which intrinsic ordering is sought between pairs of rules based on their properties such as sensitivity to derived environments, word boundaries, and so forth, thereby constituting a *cluster* of properties that, by hypothesis were ordered into relative strata. It is this latter strategy that is pursued in its application to the organization of the morphological component in Arregi and Nevins 2012.

2. There may be examples in the literature of extrinsically ordered vocabulary entries but they are usually not accepted as optimal analyses, whereas in phonology extrinsic ordering in some cases continues to be irreducible.

3. In other words, there is no need for a back-formation analysis of *self-destruct* (Aronoff 1976, 27–28), which simply receives the elsewhere allomorph. Surprising confirmation of the analysis in the text comes from the lyrics to the song "As I Destruct"

by Threat Signal (http://www.youtube.com/watch?v=zuYRul_8W9o), which employ the verb in question in an unaccusative usage, where $v*$ is not present.

4. We assume that passive *destroyed* (not **destructed*) contains transitive $v*$. This can be implemented in terms of a head Voice distinct from and higher than $v*$ that is responsible for the syntactic presence of the external argument in actives and its absence in passives, as well as other systematic differences between passive and active configurations (Kratzer 1996; Collins 2005; Pylkkänen 2008; Harley 2013).

5. Except otherwise noted, all Basque data are from our own fieldwork. In representing Basque sentences, we use orthographic conventions that are standard in the dialectological literature. We have accordingly adapted the orthography of examples whose sources do not use these conventions. For ease of exposition, all auxiliaries in the Basque examples are given in italics, and their component morphemes are separated by spaces. In addition, they are given in their surface form in isolation. Where relevant, they are followed in parentheses by the form that results from morphophonological processes that apply across word boundaries.

6. Several processes, some of which are described below, can alter the relative position of these morphemes.

7. On the absence of third-person (absolutive) proclitics, see Arregi and Nevins 2012, 52–56.

8. Furthermore, the allomorphs of T (8) and (9) are specific to auxiliaries with Ergative Metathesis or Doubling—that is, these forms are conditioned on the presence of an ergative (as opposed to absolutive) clitic in proclitic position, which shows that the proclitic is ergative even at the point of Vocabulary Insertion when the form of T is determined.

9. Ergative and Dative Doubling are not as well described as Ergative Metathesis, and it is possible that their incidence in dialectal variation in Basque verbal morphology is underreported. One indication that this might be the case is the fact that Ergative Doubling (as opposed to Metathesis) is specifically prescribed against in Batua, the standard dialect. For instance, it is listed together with other "common errors" at several points in Zubiri and Zubiri 2012 (e.g., common error #5 on p. 473 and #3 on p. 481).

10. For ease of exposition, we implement these entries in terms of informal reference to features (e.g. "first," "singular") instead of binary features such as [±participant] and [±singular]. These details are not important for the discussion.

11. Note that linearization-dependent allomorphy—where neutralization is not necessarily even at stake—is found in many languages with proclisis/enclisis alternations, such as Paduan (see Poletto 2000, 51–55; Cardinaletti and Repetti 2008; and references cited there) and Valencian Catalan (Todolí 1992).

12. In Arregi and Nevins 2012, 117–124, we propose a different analysis in which, taking advantage of the fact that Basque proclitic entries such as (11a) must make reference to the category feature of the following morpheme (T), category features are privileged over other features in determining competition at Vocabulary Insertion. Although the analysis works for Basque pronominal clitics, it does not extend to the case of Bulgarian definite articles (section 12.3 below). We would like to thank Vera Gribanova for helpful discussion of this point.

13. All Bulgarian examples are from Harizanov and Gribanova 2011, unless otherwise noted.

14. See also Sadock 1991, 117–120, for similar patterns in Macedonian, with respect to both the position of the definite article and its allomorphy.

15. We adopt a traditional structure for adjectival modification where AP is an adjunct to NP (see Dost and Gribanova 2006 for arguments specific to Bulgarian). Embick and Noyer (2001) assume a structure where NP is the complement of A (Abney 1987). The main reason for this seems to be their hypothesis that lowering can only adjoin a head to the head of its complement. However, Embick and Noyer (2001, section 7.2) relax this condition on lowering in order to account for cases where the target of this operation is clearly not the head of the complement of the lowered head. Furthermore, in the specific case of Bulgarian, the article attaches to the first adjective in examples with coordinated adjectives ((26) below), which, even under Abney's (1987) analysis, is not in any clear sense the head of the complement of D. This raises interesting questions about the workings of lowering that go well beyond the scope of the present chapter.

16. According to Bontcheva and Kilbury 2003, some animate-denoting nouns that do not end in -*o*, such as *atashe* 'attaché', seem to be exceptions, since they take the neuter allomorph -*to*. However, they trigger neuter agreement in modifying adjectives (Boris Harizanov, personal communication), which entails that they are in fact neuter (with respect to grammatical gender, which mismatches natural gender), and thus selection of the neuter allomorph of the article is expected. On the other hand, plural numerals that end in *o* (e.g., *sto* 'hundred') are genuine exceptions, since they take the stress-attracting allomorph -*té* (Scatton 1984, 171) instead of the expected -*to*. In terms of the analysis proposed below, we assume that these idiosyncratic exceptions are due to an additional vocabulary entry for -*té* contextually specified for these hosts. We would like to thank Vera Gribanova, Boris Harizanov, and Ora Matushansky for bringing these data to our attention.

17. We only provide examples of articles attaching to nouns here. See Harizanov and Gribanova 2011 for relevant examples of noun modifiers, and for further illustration of the generalizations with nouns.

18. Bulgarian neutralizes gender distinctions in the plural.

19. In the first two entries, "/-V/" is taken to mean 'ends in V'.

20. Although resolution in coordination is normally understood as operating "bottom-up" (the features on &P are determined by the features on coordinated elements), we assume that resolution rules are neutral in this respect, and can thus be used in a "top-down" fashion. Note that agreement resolution in (26)–(27) could also result in either or both conjoined adjectives being plural. The presence of singular agreement on both conjoined adjectives in (26)–(27) has a distributive effect on the meaning of these DPs, paraphrasable as 'the (friendly) Bulgarian *nation* and the (friendly) Russian *nation*' (not *nations*). It is not clear to us at this point whether this means that agreement can have semantic effects, or whether it calls for a change in the structure of coordination assumed here.

21. See below for brief comparison and discussion of differences between our account of agreement and the analyses in Arregi and Nevins 2012 and Bhatt and Walkow, forthcoming.

22. These structures abstract away from agreement relations established by D in (27) (which work the same way as in (26)) and by &P (which agrees with its sister NP). Note also that the features on &P and the conjuncts are only available after feature valuation in the postsyntactic component. We specify them in (29) for ease of exposition. Both AP and &P are phrasal probes. Although probes are normally assumed to be heads (e.g., Chomsky 2000), we assume that phrasal probes are possible (Carstens 2012). Under a traditional adjunction analysis of adjectival modification, where the agreeing adjective is inside an AP adjoined to the NP containing the agreed with noun, this assumption is necessary (at least for DP-internal agreement), unless one adopts Baker's (2008, chap. 2) proposal that agreement is possible when the goal c-commands the probe. Furthermore, in the case of coordinated APs in (29), the probe must be the plural &P, which is phrasal, not the singular As contained in it.

References

Abney, Steven. 1987. The English noun phrase in its sentential aspect. Doctoral dissertation, MIT.

Abusch, Dorit. 1997. Sequence of tense and temporal de re. *Linguistics and Philosophy* 20:1–50.

Abusch, Dorit. 2004. On the temporal composition of infinitives. In Jacqueline Guéron and Jacqueline Lecarme, eds., *The Syntax of Time*, 27–53. Cambridge, MA: MIT Press.

Ackema, Peter, and Ad Neeleman. 2003. Context-sensitive Spell-Out. *Natural Language & Linguistic Theory* 21:681–735.

Ackema, Peter, and Ad Neeleman. 2004. *Beyond Morphology: Interface Conditions on Word Formation*. Oxford: Oxford University Press.

Ackema, Peter, and Ad Neeleman. 2012. Agreement weakening at PF: A reply to Benmamoun and Lorimor. *Linguistic Inquiry* 43:75–96.

Acquaviva, Paolo. 2008. *Lexical Plurals: A Morphosemantic Approach*. Oxford: Oxford University Press.

Adger, David, and Daniel Harbour. 2007. Syntax and syncretisms of the Person Case Constraint. *Syntax* 10:2–37.

Adger, David, and Jennifer Smith. 2005. Variation and the Minimalist Program. In Leonie Cornips and Karen P. Corrigan, eds., *Syntax and Variation: Reconciling the Biological and the Social*, 149–178. Amsterdam: John Benjamins.

Akinlabi, Akinbiyi. 1996. Featural affixation. *Journal of Lingvistics* 32:239–289.

Akinlabi, Akinbiyi. 2011. Featural affixes. In Marc van Oostendorp, Colin J. Ewen, and Elizabeth V. Hume, eds., *The Blackwell Companion to Phonology*, 1945–1971. New York: Wiley.

Albizu, Pablo, and Luis Eguren. 2000. An optimality theoretic account for "ergative displacement" in Basque. In Wolfgang U. Dressler, Oskar E. Pfeiffer, Markus A. Pöchtrager, and John R. Rennison, eds., *Morphological Analysis in Comparison*, 1–23. Amsterdam: John Benjamins.

Albright, Adam, and Eric Fuß. 2012. Syncretism. In Jochen Trommer, ed., *The Morphology and Phonology of Exponence*, 236–288. Oxford: Oxford University Press.

Alcoba, Santiago. 1999. La flexión verbal. In Ignacio Bosque and Violeta Demonte, eds., *Gramática descriptiva de la lengua española*, 4915–4992. Madrid: Espasa.

Allen, Barbara, Donna Gardiner, and Donald Frantz. 1984. Noun incorporation in Southern Tiwa. *International Journal of American Linguistics* 50:292–311.

Alsina, Àlex. 2002. L'infinitiu. In Joan Solà, Maria Rosa Lloret, Joan Mascaró, and Manuel Pérez Saldanya, eds., *Gramàtica del català contemporani*, 2389–2454. Barcelona: Empúries.

Amaya, María. 1999. *Damana*. Munich: Lincom Europa.

Anagnostopoulou, Elena, and Yota Samioti. Forthcoming. *Domains within words and their meanings: A case study*. In Artemis Alexiadou, Hagit Borer, and Florian Schäfer, eds., *The Syntax of Roots and the Roots of Syntax*. Oxford: Oxford University Press.

Anderson, Stephen. 1984. On representations in morphology: Case marking, agreement and inversion in Georgian. *Natural Language & Linguistic Theory* 2:157–218.

Anderson, Stephen. 1992. *A-Morphous Morphology*. Cambridge: Cambridge University Press.

Anderson, Stephen. 2008. Phonologically conditioned allomorphy in the morphology of Surmiran (Rumantsch). In Angela Ralli, Geert Booij, Sergio Scalise, and Athanasios Karasimos, eds., *On-line Proceedings of the 6th Mediterranean Morphology Meeting (MMM6)*, 32–48. Ithaca, Greece: University of Patras, Universiteit Leiden, Università degli Studi di Bologna. http://morbo.lingue.unibo.it/mmm.

Anttila, Arto. 1997. Deriving variation from grammar. In Frans Hinskens, Roeland van Hout, and W. Leo Wetzels, eds., *Variation, Change and Phonological Theory*, 35–68. Amsterdam: John Benjamins.

Aoun, Joseph, Elabbas Benmamoun, and Dominique Sportiche. 1994. Agreement, word order, and conjunction in some varieties of Arabic. *Linguistic Inquiry* 25:195–220.

Arad, Maya. 2003. Locality constraints on the interpretation of roots: The case of Hebrew denominal verbs. *Natural Language & Linguistic Theory* 21:737–778.

Aronoff, Mark. 1974. Word-structure. Doctoral dissertation, MIT.

Aronoff, Mark. 1976. *Word Formation in Generative Grammar*. Cambridge, MA: MIT Press.

Aronoff, Mark. 1988. Head operations and strata in reduplication: A linear treatement. *Yearbook of Morphology* 1:1–15.

Aronoff, Mark. 1994. *Morphology by Itself: Stems and Inflectional Classes*. Cambridge, MA: MIT Press.

Aronoff, Mark, and Kirsten Fudeman. 2011. *What is Morphology?* 2nd ed., Chichester:Wiley-Blackwell.

Aronson, Howard I. 1990. *Georgian: A Reading Grammar*. Rev. ed. Columbus, OH: Slavica.

Arregi, Karlos. 2000. Tense in Basque. Ms., MIT.

Arregi, Karlos. 2001. Person and number inflection in Basque. In Beatriz Fernández and Pablo Albizu, eds., *On Case and Agreement*, 71–111. Bilbo: Euskal Herriko Unibertsitatea.

Arregi, Karlos, and Andrew Nevins. 2012. *Morphotactics: Basque Auxiliaries and the Structure of Spellout*. Dordrecht: Springer.

Badia i Margarit, Antoni M. 1951. *Gramática histórica catalana*. (Cited from the first Catalan edition, *Gramàtica històrica catalana*, València: Tres i Quatre, 1981.)

Badia i Margarit, Antoni M. 1994. *Gramàtica de la llengua catalana*. Barcelona: Enciclopèdia Catalana.

Baerman, Matthew. 2007. Morphological reversals. *Journal of Linguistics* 43:33–61.

Baker, Mark. 1988. *Incorporation: A Theory of Grammatical Function Changing*. Chicago: University of Chicago Press.

Baker, Mark. 2008. *The Syntax of Agreement and Concord*. Cambridge: Cambridge University Press.

Banksira, Degif Petros. 2000. *Sound Mutations: The Morphophonology of Chaha*. Amsterdam: John Benjamins.

Bartsch, Renate. 1987. The construction of properties under perspectives. *Journal of Semantics* 5:293–320.

Bayer, Josef. 1996. *Directionality and Logical Form*. Dordrecht: Kluwer Academic Publishers.

Beard, Robert. 1995. *Lexeme-Morpheme Based Morphology: A General Theory of Inflection and Word Formation*. Albany: State University of New York Press.

Beck, Sigrid. 2000. The semantics of *different*: Comparison operator and relational adjective. *Linguistics and Philosophy* 23:101–139.

Béjar, Susana. 2003. Phi-syntax: A theory of agreement. Doctoral dissertation, University of Toronto.

Béjar, Susana. 2008. Conditions on Phi-Agree. In Daniel Harbour, David Adger, and Susana Béjar, eds., Phi Theory: Phi Features across Modules and Interfaces, 130–154. Oxford: Oxford University Press.

Béjar, Susana, and Daniel Hall. 1999. Marking markedness: The underlying order of diagonal syncretisms. Paper presented at the Eastern States Conference on Linguistics (ESCOL), University of Connecticut, Storrs, 1999.

Béjar, Susana, and Milan Rezac. 2003a. Cyclic agree. Toronto Working Papers in Linguistics, University of Toronto.

Béjar, Susana, and Milan Rezac. 2003b. Person licensing and the derivation of PCC effects. In Ana Teresa Pérez-Leroux and Yves Roberge, eds., *Romance Linguistics*, 49–62. Amsterdam: John Benjamins.

De Belder, Marijke. 2011. *Roots and Affixes: Eliminating Lexical Categories from Syntax*. Utrecht: LOT.

Bermúdez-Otero, Ricardo. 2006. Morphological structure and phonological domains in Spanish denominal derivation. In Fernando Martínez-Gil and Sonia Colina, eds., *Optimality-Theoretic Studies in Spanish Phonology*, 278–311. Amsterdam: John Benjamins.

Bermúdez-Otero, Ricardo. 2013. The Spanish lexicon stores stems with theme vowels, not roots with inflectional class features. *Probus* 25:1.

Bhatt, Rajesh. 2000. Covert modality in non-finite contexts. Doctoral dissertation, University of Pennsylvania.

Bhatt, Rajesh, and Roumyana Pancheva. 2004. Late merge of degree clauses. *Linguistic Inquiry* 35:1–45.

Bhatt, Rajesh, and Martin Walkow. Forthcoming. Locating agreement in grammar: An argument from agreement in conjunctions. *Natural Language & Linguistic Theory*.

Bierwisch, Manfred. 1989. The semantics of gradation. In Manfred Bierwisch and Ewald Lang, eds., *Dimensional Adjectives*, 71–261. Berlin: Springer-Verlag.

Blair, Robert W., and Refugio Vermont Salas. 1995. *Spoken Yucatec Maya.* Chapel Hill, NC: Duke-UNC Consortium in Latin American Studies.

Bliss, Heather. 2005. Formalizing point-of-view: The role of sentience in Blackfoot's direct/inverse system. MA thesis, University of Calgary.

Bobaljik, Jonathan. 2000. The ins and outs of contextual allomorphy. In Kleanthes K. Grohmann and Caro Struijke, eds., *University of Maryland Working Papers in Linguistics* 10, 35–71. College Park: University of Maryland.

Bobaljik, Jonathan. 2002. Syncretism without paradigms: Remarks on Williams 1981, 1994. In Geert Booij and Jaap van Marle, eds., *Yearbook of Morphology 2001*, Dordrecht: Kluwer.

Bobaljik, Jonathan. 2008. Paradigms (Optimal and otherwise): A case for skepticism. In Asaf Bachrach and Andrew Nevins, eds., *Inflectional Identity*, 29–54. Oxford: Oxford University Press.

Bobaljik, Jonathan. 2012. *Universals in Comparative Morphology: Suppletion, Superlatives, and the Structure of Words.* Cambridge, MA: MIT Press.

Bobaljik, Jonathan, and Susi Wurmbrand. 2005. The domain of agreement. *Natural Language & Linguistic Theory* 23:809–865.

Bonet, Eulàlia. 1991. Morphology after syntax: Pronominal clitics in Romance. Doctoral dissertation, MIT.

Bonet, Eulàlia, and Daniel Harbour. 2012. Contextual allomorphy. In Jochen Trommer, ed., *The Morphology and Phonology of Exponence*, 195–235. Studies in Theoretical Linguistics. Oxford: Oxford University Press.

Bonet, Eulàlia, Maria-Rosa Lloret, and Joan Mascaró. Forthcoming. The prenominal allomorphy syndrome. In Eulàlia Bonet, Maria-Rosa Lioret, and Joan Mascaró, eds., *Understanding Allomorphy: Perspectives from Optimality Theory*. London: Equinox.

Bonet, Eulàlia, and Joan Mascaró. 2012. Asimetrías de concordancia en el SD: El rasgo de masa en asturiano. In Antonio Fábregas, Elena Felíu, Josefa Martín, and José Pazó, eds., *Los límites de la morfología: Estudios ofrecidos a Soledad Varela Ortega*, 91–104. Madrid: Servicio de Publicaciones de la Universidad Autónoma de Madrid.

Bontcheva, Katina, and James Kilbury. 2003. An inheritance-based description of Bulgarian noun inflection. In S. Piperides and V. Karkaletsis, eds., *International Workshop "Balkan Language Resources and Tools" in the Framework of the 1st Balkan Conference in Informatics (BCI)*, 23–27. Seville, Spain: Incoma.

Bošković, Željko. 1996. Selection and the categorial status of Infinitival Complements. *Natural Language & Linguistic Theory* 14:269–304.

Bošković, Željko. 1997. *The Syntax of Nonfinite Complementation: An Economy Approach.* Cambridge, MA: MIT Press.

Bošković, Željko. 2002. A-Movement and the EPP. *Syntax* 5:167–218.

Bošković, Željko. 2007. On the locality and motivation of Move and Agree: An even more minimal theory. *Linguistic Inquiry* 38:589–644.

Bošković, Željko. 2010. Phases beyond clauses. Ms., University of Connecticut, Storrs.

Bouchard, Denis. 2002. *Adjectives, Number and Interfaces: Why Languages Vary.* Oxford: Elsevier Science.

Bouchard, Denis. 2005. Sériation des adjectifs dans le SN et formation de concepts. In Patricia Cabredo Hofherr and Ora Matushansky, eds., *L'Adjectif,* 125–142. Recherches Linguistiques de Vincennes 34. Saint-Denis: Presses Universitaires de Vincennes.

Bowers, John. 1975. Adjectives and adverbs in English. *Foundations of Language* 13:529–562.

Bowers, John. 1987. Extended X-bar theory, the ECP and the Left Branch Condition. In Megan Crowhurst, ed., *Proceedings of WCCFL 6,* 47–62. stanford, CA: CSLI.

Bresnan, Joan. 1973. Syntax of the comparative clause construction in English. *Linguistic Inquiry* 4:275–343.

Brody, Michael. 1997. Perfect chains. In Liliane Haegeman, ed., *Elements of Grammar: Handbook of Generative Syntax,* 139–167. Dordrecht: Kluwer Academic Publishers.

Brownie, John, and Marjo Brownie. 2007. *Mussau Grammar Essentials.* Ukarumpa, Papua New Guinea: SIL-PNG Academic Publications. www.sil.org/pacific/png/pubs/48552/MussauGrammarEssentials.pdf.

Bruening, Benjamin. 2001. Syntax at the edge: Cross-clausal phenomena and the syntax of Passamaquoddy. Doctoral dissertation, MIT.

Burzio, Luigi. 1994. *Principles of English Stress.* Cambridge: Cambridge University Press.

Burzio, Luigi. 2007. Phonology and phonetics of English stress and vowel reduction. *Language Sciences* 29:154–176.

Bye, Patrik, and Peter Svenonius. 2012. Non-concatenative morphology as epiphenomenon. In Jochen Trommer, ed., *The Morphology and Phonology of Exponence,* 427–494. Oxford: Oxford University Press.

Caha, Pavel. 2009. The nanosyntax of case. Docforal dissertation, CASTL Tromsø.

Calabrese, Andrea. 1999. Metaphony revisited. *Rivista di Linguistica* 10:7–68.

Calabrese, Andrea. 2009. Metaphony. Ms., University of Connecticut.

Calabrese, Andrea. 2012. Allomorphy in the Italian Passato Remoto: A Distributed Morphology analysis. Ms., University of Connecticut.

Cardinaletti, Anna, and Giuliana Giusti. 2001. "Semi-lexical" motion verbs in Romance and Germanic. In Norbert Corver and Henk van Riemsdijk, eds., *Semi-lexical Categories. On the Function of Content Words and the Content of Function Words,* 371–414. Berlin: Mouton de Gruyter.

Cardinaletti, Anna, and Lori Repetti. 2008. The phonology and syntax of preverbal and postverbal subject clitics in Northern Italian dialects. *Linguistic Inquiry* 39:523–563.

Carlson, Gregory Norman. 1977. A unified analysis of the English bare plural. *Linguistics and Philosophy* 1:413–457.

Carstairs, Andrew. 1987. *Allomorphy in Inflection*. Croom Helm Linguistics Series. London: Croom Helm.

Carstairs-McCarthy, Andrew. 1992. *Current Morphology*. London: Routledge.

Carstens, Vicki. 2012. Delayed valuation: A reanalysis of "upwards" complementizer agreement and the mechanics of case. Ms., University of Missouri, Columbia.

Castillo, Juan Carlos, John E. Drury, and Kleanthes K. Grohmann. 2009. Merge Over Move and the Extended Projection Principle: MOM and the EPP Revisited. *Iberia: An International Journal of Theoretical Linguistics* 1 (1):53–114.

Cecchetto, Carlo. 2003. QR in the theory of phases. In Gina Garding and Mimu Tsujimura, eds., Proceedings of the 22nd West Coast Conference on Formal Linguistics (WCCFL), 123–136. Somerville, MA: Cascadilla Press.

Cecchetto, Carlo. 2004. Explaining the locality conditions of QR: Consequences for the theory of phases. *Natural Language Semantics* 12:345–397.

Chomsky, Noam. 1957. *Syntactic Structures*. The Hague: Mouton.

Chomsky, Noam. 1973. Conditions on transformations. In Stephen R. Anderson and Paul Kiparsky, eds., *A Festschrift for Morris Halle*, 232–286. New York: Holt, Rinehart and Winston.

Chomsky, Noam. 1986. *Barriers*. Cambridge MA: MIT Press.

Chomsky, Noam. 1995. *The Minimalist Program*. Cambridge, MA: MIT Press.

Chomsky, Noam. 2000. Minimalist inquiries: The framework. In Roger Martin, David Michaels, and Juan Uriagereka, eds., *Step by Step: Essays on Minimalist Syntax in Honor of Howard Lasnik*, 89–156. Cambridge, MA: MIT Press.

Chomsky, Noam. 2001. Derivation by phase. In Michael Kenstowicz, ed., *Ken Hale: A Life in Language*, 1–52. Cambridge, MA: MIT Press.

Chomsky, Noam, and Morris Halle. 1968. *The Sound Pattern of English*. New York: Harper and Row.

Chung, Inkie. 2007. Suppletive negation in Korean and Distributed Morphology. *Lingua* 117:95–148.

Cinque, Guglielmo. 2005. Deriving Greenberg's Universal 20 and its exceptions. *Linguistic Inquiry* 36:315–332.

Cinque, Guglielmo. 2010. *The Syntax of Adjectives: A Comparative Study*. Cambridge, MA: MIT Press.

Cipria, Alicia, and Craige Roberts. 2001. Spanish imperfecto and pretérito: Truth conditions and Aktionsart effects in a Situation Semantics. Revised version of a paper by the same name that appeared in 1996, in Jae-Hak Yoon and Andreas Kathol, eds., *Papers in Semantics: OSU Working Papers in Linguistics 49*, 43–70. Columbus: Ohio State University.

Clarke, John. 2001. Children's production of comparative *more* and *-er*: Elicited production versus relative judgement. In Florencia Franceschina, Michelle Kingston, and Emma Thomas, eds., *Essex Graduate Student Papers in Language and Linguistics*, vol. 4, 1–18. Colchester: Department of Language and Linguistics, University of Essex.

Clua, Esteve. 1999. Variació i distància lingüística: Classificació dialectal del valencià a partir de la morfologia flexiva. Doctoral dissertation, Universitat de Barcelona.

Collins, Chris. 2005. A smuggling approach to the passive in English. *Syntax* 8:81–120.

Colón, Germà. 1976. Sobre el perfet perifràstic *vado*+infinitiu en català, en provençal i en francès. In Felip M. Lorda i Alaiz and Jean Roudil, eds., *Problemes de llengua i literatura catalanes: Actes del II Col·loqui Internacional sobre el català, Amsterdam 1970*, 101–144. Barcelona: Publicacions de l'Abadia de Montserrat.

Comajoan, Llorenç, and Manuel Pérez Saldanya. 2005. Grammaticalization and language acquisition: Interaction of lexical aspect and discourse. In David Eddington, ed., *Selected Proceedings of the 6th Conference on the Acquisition of Spanish and Portuguese as First and Second Languages*, 44–55. Somerville, MA: Cascadilla Proceedings Project.

Corbett, Greville. 1983. *Hierarchies, Targets and Controllers: Agreement Patterns in Slavic*. London: Croom Helm.

Corbett, Greville. 2000. *Number*. Cambridge: Cambridge University Press.

Corbett, Greville. 2010. Features: Essential notions. In Anna Kibort and Greville Corbett, eds., *Features*, 17–36. Oxford: Oxford University Press.

Corver, Norbert. 1990. The syntax of left branch extractions. Doctoral dissertation, University of Tilburg.

Corver, Norbert. 1991. Evidence for DegP. In Tim Sherer, ed., *Proceedings of NELS 21*, 33–47. Amherst: GLSA, University of Massachusetts.

Corver, Norbert. 1997a. The internal syntax of the Dutch extended adjectival projection. *Natural Language & Linguistic Theory* 15:289–368.

Corver, Norbert. 1997b. *Much*-support as a last resort. *Linguistic Inquiry* 28:119–164.

Corver, Norbert. 2005. Double comparatives and the comparative criterion. In Patricia Cabredo Hofherr and Ora Matushansky, eds., *L'Adjectif*, 165–190. Recherches Linguistiques de Vincennes 34. Saint-Denis: Presses Universitaires de Vincennes.

Cowper, Elizabeth. 2005. A note on number. *Linguistic Inquiry* 36:441–455.

Curnow, Timothy. 2000. Why "first/non-first person" is not grammaticalized mirativity. In Keith Allan and John Henderson, eds., *Proceedings of ALS2k: The 2000 Conference of the Australian Linguistic Society*. http:: //www.als.asn.au.

Curnow, Timothy. 2002. Conjunct/disjunct marking in Awa Pit. *Linguistics* 40:611–627.

Cutillas, Juan Antonio. 2003. *Teoría lingüística de la optimidad: Fonología, morfología y aprendizaje*. Murcia: Servicio de Publicaciones de la Universidad de Murcia.

Davies, Mark. 2008–. The Corpus of Contemporary American English: 425 million words, 1990–present. http://corpus.byu.edu/coca/.

Den Dikken, Marcel. 2007. Phase extension: Contours of a theory of the role of head movement in phrasal extraction. *Theoretical Linguistics* 33:1–41.

DiSciullo, Anna-Maria, and Edwin Williams. 1987. *On the Definition of Word.* Cambridge, MA: MIT Press.

Dixon, R. M. W. 1983. Nyawaygi. In R. M. W. Dixon and Barry J. Blake, eds., *Handbook of Australian Languages*, vol. 3, 431–525. Amsterdam: John Benjamins.

Doetjes, Jenny, Ad Neeleman, and Hans van de Koot. 1998. Degree expressions and the autonomy of syntax. *UCL Working Papers in Linguistics* 10:323–368.

Dost, Ascander, and Vera Gribanova. 2006. Definiteness marking in the Bulgarian. In Donald Baumer, David Montero, and Michael Scanlon, eds., *Proceedings of the 25th West Coast Conference on Formal Linguistics (WCCFL)*, 132–140. Somerville, MA: Cascadilla Press.

Van Driem, George. 1987. *A Grammar of Limbu.* Berlin: Mouton de Gruyter.

Dumézil, George. 1931. *La langue des oubykhs.* Paris: Honoré Champion.

Eddington, David, and José Ignacio Hualde. 2008. El abundante agua fría: Hermaphroditic Spanish nouns. *Studies in Hispanic and Lusophone Linguistics* 1:5–31.

Ekdahl, Muriel, and Joseph E. Grimes. 1964. Terena verb inflection. *International Journal of American Linguistics* 30:261–268.

Embick, David. 1997. Voice and the interfaces of syntax. Doctoral dissertation, University of Pennsylvania.

Embick, David. 1998. Voice systems and the syntax/morphology interface. In Heidi Harley, ed., *MITWPL 32: Papers from the UPenn/MIT Roundtable on Argument Structure and Aspect*, 41–72. Cambridge, MA: MITWPL, Department of Linguistics and Philosophy, MIT.

Embick, David. 2000. Features, syntax and categories in the Latin perfect. *Linguistic Inquiry* 31:185–230.

Embick, David. 2003. Locality, listedness, and morphological identity. *Studia Linguistica* 57:143–169.

Embick, David. 2007a. Blocking effects and analytic/synthetic alternations. *Natural Language & Linguistic Theory* 25:1–37.

Embick, David. 2007b. Linearization and local dislocation: Derivational mechanics and interactions. *Linguistic Analysis* 33:303–336.

Embick, David. 2008. Variation and morphosyntactic theory: Competition fractionated. *Language and Linguistics Compass* 2:59–78.

Embick, David. 2010. *Localism versus Globalism in Morphology and Phonology.* Cambridge, MA: MIT Press.

Embick, David. 2012. Contextual conditions on stem alternations: Illustrations from the Spanish conjugation. In Irene Franco, Sara Lusini, and Andrés Saab, eds., *Romance Languages and Linguistic Theory 2010: Selected Papers from "Going Romance," Feidem 2010.* 21–40. Amsterdam: John Benjamins.

Embick, David, and Morris Halle. 2005. On the status of *stems* in morphological theory. In T. Geerts and H. Jacobs, eds., *Proceedings of Going Romance 2003*, 59–88. Amsterdam: John Benjamins.

Embick, David, and Alec Marantz. 2005. Cognitive neuroscience and the English past tense: Comments on the paper by Ullman et al. *Brain and Language* 93:243–247.

Embick, David, and Alec Marantz. 2008. Architecture and blocking. *Linguistic Inquiry* 39:1–53.

Embick, David, and Rolf Noyer. 1999. Locality in post-syntactic operations. In Vivian Lin, Cornelia Krause, Benjamin Bruening, and Karlos Arregi, eds., *MITWPL 34: Papers in Morphology and Syntax*, 265–317. Cambridge, MA: MITWPL, Department of Linguistics and Philosophy, MIT.

Embick, David, and Rolf Noyer. 2001. Movement operations after syntax. *Linguistic Inquiry* 32:555–595.

Embick, David, and Rolf Noyer. 2007. Distributed Morphology and the Syntax/Morphology Interface. In Gillian Ramchand and Charles Reiss, eds., *The Oxford Handbook of Linguistic Interfaces*, 289–324. Oxford: Oxford University Press.

Emonds, Joseph. 1976. *A Transformational Approach to English Syntax: Root, Structure-Preserving, and Local Transformations*. New York: Academic Press.

Enç, Mürvet. 2004. Rethinking past tense. In Jacqueline Guéron and Jacqueline Lecarme, eds., *The Syntax of Time*, 203–215. Cambridge, MA: MIT Press.

England, Nora. 1983. *A Grammar of Mam, a Mayan Language*. Austin: University of Texas Press.

Evans, Nicholas. 2003. *Bininj Gun-Wok: A Pan-Dialectal Grammar of Mayali, Kunwinjku and Kune*. Canberra: Pacific Linguistics.

Fernández, Beatriz. 2001. Absolutibo komunztaduradun ergatiboak, absolutibo komunztaduradun datiboak: Ergatiboaren lekualdatzetik datiboaren lekualdatzera. In Beatriz Fernández and Pablo Albizu, eds., *Kasu eta Komunztaduraren Gainean: On Case and Agreement*, 147–165. Bilbo: Euskal Herriko Unibertsitatea. http://basdisyn. net/pdf/Kasu eta komunztaduraren%20gainean.pdf.

Fernández, Beatriz, and Pablo Albizu. 2000. Ergative displacement in Basque and the division of labor between morphology and syntax. In Akira Okrent and John P. Boyle, eds., *Proceedings of CLS 36, Vol. 2: The Panels*. 103–117. Chicago: Chicago Linguistic Society, University of Chicago.

Fernández, Beatriz, and María José Ezeizabarrena. 2003. Itsasaldeko solezismoa, datiboaren lekualdatzearen argipean. In Jesus Mari Makazaga and Bernard Oyharçabal, eds., *Euskaltzaindiaren Biltzarra 15*, 255–277. Bilbo: Euskaltzaindia. http://www. euskaltzaindia.org/dok/ikerbilduma/54817.pdf.

Fernández-Ordóñez, Inés. 2007. El 'neutro de materia' en Asturias y Cantabria: Análisis gramatical y nuevos datos. In Inmaculada Delgados Cobos and Alicia Puigvert Ocal, eds., *Ex admiratione et amicitia: Homenaje a Ramón Santiago*, 395–434. Madrid: Ediciones del Orto.

Fitzpatrick, Justin. 2002. On minimalist approaches to the locality of movement. *Linguistic Inquiry* 33:443–463.

Foley, William. 1986. *The Papuan Languages of New Guinea*. Cambridge: Cambridge University Press.

Fortson, Benjamin W. 2010. *Indo-European Language and Culture: An Introduction*. Malden, MA: Wiley-Blackwell.

Fox, Danny. 1999. Reconstruction, binding theory, and the interpretation of chains. *Linguistic Inquiry* 30:157–196.

Fox, Danny. 2000. *Economy and Semantic Interpretation*. Cambridge, MA: MIT Press/ MITWPL.

Franck, Julie, Glenda Lassi, Ulrich H. Frauenfelder, and Luigi Rizzi. 2006. Agreement and movement: A syntactic analysis of attraction. *Cognition* 101:173–216.

Franks, Steven. 2001. The internal structure of Slavic NPs, with special reference to Bulgarian. In Adam Przepiorkowski and Piotr Banski, eds., *Generative Linguistics in Poland (GLiP) 2*, 53–69. Warsaw: Polskiej Akademii Nauk, Instytut Podstaw Informatyki.

Fudge, Eric. 1984. *English Word-Stress*. London: Allen and Unwin.

Fuß, Eric. 2012. Syncretism. In Jochen Trommer, ed., *The Morphology and Phonology of Exponence*, 236–288. Oxford: Oxford University Press.

Gallego, Ángel. 2005. Phase sliding. Ms., University of Barcelona and University of Maryland, College Park.

Gallego, Ángel. 2010. Phase theory. Amsterdam: John Benjamins.

Gallego, Ángel, and Juan Uriagereka. 2006. Sub-extraction from subjects. Ms., University of Barcelona and University of Maryland, College Park.

Gavarró, Anna, and Brenda Laca. 2002. Les perífrasis temporals, aspectuals i modals. In Joan Solà, Maria Rosa Lloret, Joan Mascaró, and Manuel Pérez Saldanya, eds., *Gramàtica del català contemporani*, 2663–2726. Barcelona: Empúries.

Giorgi, Alessandra, and Fabio Pianesi. 1998. *Tense and Aspect: From Semantics to Morphosyntax*. Oxford: Oxford University Press.

Gómez Durán, Gemma. 2011. Gramàtica del català rossellonès. Doctoral dissertation, Universitat Autònoma de Barcelona.

Gómez Torrego, Leonardo. 1999. Los verbos auxiliares: Las perífrasis verbales de infinitivo. In Ignacio Bosque and Violeta Demonte, eds., *Gramática descriptiva de la lengua española*, 3323–3390. Madrid: Espasa.

González-Díaz, Victorina. 2007. On the nature and distribution of English double periphrastic comparison. *Review of English Studies* 57:623–664.

González-Poot, Antonio, and Martha McGinnis. 2006. Local versus long-distance fission in Distributed Morphology. In Claire Gurski, ed., *Proceedings of the 2005 Annual Conference of the Canadian Linguistic Association*. http://westernlinguistics. ca/ Publications/CLA-ACL/CLA-ACL2005.htm.

Graziano-King, Janine. 1999. Acquisition of comparative forms in English. Doctoral dissertation, CUNY.

Grohmann, Kleanthes K, ed. 2009. *InterPhases: Phase-Theoretic Investigations of Linguistic Interfaces*. Oxford: Oxford University Press.

Grohmann, Kleanthes K., John Drury, and Juan Carlos Castillo. 2000. No more EPP. In Roger Billerey and Brook Danielle Lillehaugen, eds., *Proceedings of the 19th West Coast Conference on Formal Linguistics (WCCFL)*, 153–166. Somerville, MA: Cascadilla Press.

Guasti, Maria Teresa, and Luigi Rizzi. 2002. Agreement and tense as distinct syntactic positions: Evidence from acquisition. In Guglielmo Cinque, ed., *The Structure of DP and IP—The Cartography of Syntactic Structures,* vol. 1, 167–194. Oxford: Oxford University Press.

Hagman, Roy S. 1977. *Nama Hottentot Grammar.* Bloomington: Indiana University Press.

Haiman, John, and Paola Benincà. 1992. *The Raetho-Romance Languages.* London: Routledge.

Hale, Kenneth. 1997. Some observations on the contribution of local languages to linguistic science. *Lingua* 100:71–89.

Halle, Morris. 1957. In defence of the number two. In Ernst Pulgram, ed., *Studies Presented to Joshua Whatmough on His Sixtieth Birthday*, 65–72. The Hague: Mouton.

Halle, Morris. 1959. *The Sound Pattern of Russian.* The Hague: Mouton.

Halle, Morris 1973. Prolegomena to a theory of word formation. *Linguistic Inquiry* 4 (1): 3–16.

Halle, Morris. 1990. An approach to morphology. In Juli Carter, Rose-Marie Dechaine, Bill Philip, and Tim Sherer, eds., *Proceedings of NELS 20*, 150–184. Amherst: GLSA, University of Massachusetts.

Halle, Morris. 1995. Feature geometry and feature spreading. *Linguistic Inquiry* 26:1–46. Reprinted in Morris Halle, ed., *From Memory to Speech and Back: Papers on Phonetics and Phonology 1954–2002*, 196–252. Berlin: Mouton de Gruyter, 2002.

Halle, Morris. 1997a. Distributed Morphology: Impoverishment and Fission. In Benjamin Bruening, Yoonjung Kang, and Martha McGinnis, eds., *MITWPL 30: Papers at the Interface*, 425–449. Cambridge, MA: MITWPL, Department of Linguistics and Philosophy, MIT.

Halle, Morris. 1997b. On stress and accent in Indo-European. *Language* 73:275–313.

Halle, Morris. 1998. The stress of English words 1968–1998. *Linguistic Inquiry* 29:539–568.

Halle, Morris. 2001. Infixation versus onset metathesis in Tagalog, Chamorro, and Toba Batak. In Michael Kenstowicz, ed., *Ken Hale: A Life in Language*, 153–168. Cambridge: MA: MIT Press.

Halle, Morris, James W. Harris, and Jean-Roger Vergnaud. 1991. A reexamination of the Stress Erasure Convention and Spanish Stress. *Linguistic Inquiry* 22:141–159.

Halle, Morris, and William Idsardi. 1995. General properties of stress and metrical structure. In J. Goldsmith, ed., *The Handbook of Phonological Theory*, 402–443. Oxford: Blackwell.

Halle, Morris, and Michael Kenstowicz. 1991. The Free Element Condition and cyclic versus noncyclic stress. *Linguistic Inquiry* 22:457–501.

Halle, Morris, and Paul Kiparsky. 1977. Towards a reconstruction of the Proto-Indo-European accent. In Larry M. Hyman, ed., *Studies in Stress and Accent,* 209–238. Los Angeles: USC Department of Linguistics.

Halle, Morris, and Alec Marantz. 1993. Distributed Morphology and the pieces of inflection. In Kenneth Hale and Samuel J. Keyser, eds., *The View from Building 20: Essays in Linguistics in Honor of Sylvain Bromberger*, 111–176. Cambridge, MA: MIT Press.

Halle, Morris, and Alec Marantz. 1994. Some key features of Distributed Morphology. In Andrew Carnie, Heidi Harley, and Tony Bures, eds., *MITWPL 21: Papers on Phonology and Morphology*, 275–288. Cambridge, MA: MITWPL, Department of Linguistics and Philosophy, MIT.

Hale, Morris, and Jean-Roger Vergnaud. 1987. *An Essay on Stress*. Cambridge, MA: MIT Press.

Hanson, Rebecca. 2003. Why can't we all just agree?: Animacy and the person case constraint. MA thesis, University of Calgary.

Harbour, Daniel. 2007. *Morphosemantic Number: From Kiowa Noun Classes to UG Number Features*. Dordrecht: Springer.

Harbour, Daniel. 2011a. Descriptive and explanatory markedness. *Morphology* 21:223–240.

Harbour, Daniel. 2011b. The generative syntax of the natural numbers. Ms., Queen Mary University of London.

Harbour, Daniel. 2011c. Paucity, abundance, and the theory of number. Ms., Queen Mary University of London.

Harbour, Daniel. 2011d. Valence and atomic number. *Linguistic Inquiry* 42:561–594.

Harbour, Daniel. 2012. The structure of persons and the atoms of thought. Ms., Queen Mary University of London.

Harbour, Daniel, and Christian Elsholtz. 2011. Feature geometry: Self-destructed. Ms., Queen Mary University of London and Technische Universität Graz.

Harizanov, Boris, and Vera Gribanova. 2011. The role of morphological and phonological factors in Bulgarian allomorph selection. In Nick LaCara, Anie Thompson, and Matt A. Tucker, eds., *Morphology at Santa Cruz: Papers in Honor of Jorge Hankamer*, 31–40. Santa Cruz: Linguistics Research Center, University of California.

Harizanov, Boris, and Vera Gribanova. 2012. Inward sensitive contextual allomorphy and its conditioning factors. Paper presented at the North East Linguistic Society, City University of New York, October 19–21. http://people.ucsc.edu/˜bharizan/presentations.html.

Harley, Heidi. 1994. Hug a tree: Deriving the morphosyntactic feature hierarchy. In Andrew Carnie, Heidi Harley, and Tony Bures, eds., *MITWPL 21: Papers on Phonology and Morphology*, 289–320. Cambridge, MA: MITWPL, Department of Linguistics and Philosophy, MIT.

Harley, Heidi. 2010. Affixation and the Mirror Principle. In Raffaella Folli and Christiane Ullbricht, eds., *Interfaces in Linguistics*, 166–186. Oxford: Oxford University Press.

Harley, Heidi. 2011. On the identity of roots. Paper presented at the Approaches to the Lexicon Workshop (Roots III), Hebrew University of Jerusalem, Israel, June 13.

Harley, Heidi. 2013. External arguments and the Mirror Principle: On the distinctness of Voice and v. *Lingua* 125:34–57.

Harley, Heidi, and Rolf Noyer. 2000. Licensing in the non-lexicalist lexicon. In Bert Peeters, ed., *The Lexicon/Encyclopaedia Interface*, 349–374. Amsterdam: Elsevier Press.

Harley, Heidi, and Elizabeth Ritter. 2002. Person and number in pronouns: A feature-geometric analysis. *Language* 78:482–526.

Harley, Heidi, and Mercedes Tubino Blanco. 2010. Dos tipos de base verbal en Hiaki (yaqui). In Rosa María Ortiz Ciscomani, ed., *Análisi lingüístico: enfoques sincrónico, diacrónico e interdisciplinario*, 97–128. Hermosillo, Son., Mexico: Editorial Unison.

Harley, Heidi, and Mercedes Tubino Blanco. 2012. Phase-theoretic mismatches in the morphology and syntax of Hiaki verb stems. Talk presented at *Exploring the interfaces 1: Word Structure*, held at McGill University, May 6–8.

Harley, Heidi, Mercedes Tubino Blanco, and Jason Haugen. 2009. Applicative constructions and suppletive verbs in Hiaki (Yaqui). In L. Lanz, A. Franklin, J. Hoecker, E. Gentry Brunner, M. Morrison, and C. Pace, eds., *Rice Working Papers in Linguistics* 1:42–51.

Harrington, John P. 1946. Three Kiowa texts. *International Journal of American Linguistics* 12:237–242.

Harris, Alice. 1981. *Georgian Syntax: A Study in Relational Grammar*. Cambridge: Cambridge University Press.

Harris, James W. 1969. *Spanish Phonology*. Cambridge, MA: MIT Press.

Harris, James W. 1985. Spanish diphthongization and stress: A paradox resolved. *Phonology Yearbook* 2:31–45.

Harris, James W. 1987. Disagreement rules, referral rules and the Spanish feminine article *el*. *Journal of Linguistics* 23:177–183.

Harris, James. 1996. The syntax and morphology of class marker suppression in Spanish. In Karen Zagona, ed., *Grammatical Theory and Romance Languages*, 99–122. Amsterdam: John Benjamins.

Harris, James, and Morris Halle. 2005. Unexpected plural inflections in Spanish: Reduplication and metathesis. *Linguistic Inquiry* 36:192–222.

Haugen, Jason. 2008. *Morphology at the Interfaces: Reduplication and Noun Incorporation in Uto-Aztecan*. Amsterdam: John Benjamins.

Hay, Jennifer, and Harald Baayen. 2005. Shifting paradigms: Gradient structure in morphology. *Trends in Cognitive Sciences* 9:342–348.

Hayes, Bruce. 1985. *A Metrical Theory of Stress Rules*. New York: Garland.

Heath, Jeffrey. 1976. Antipassivization: A functional typology. In Henry Thompson, Kenneth Whistler, Vicky Edge, Jeri J. Jaeger, Ronya Javkin, Miriam Petruck, Christopher Smeall, and Robert D. Van Valin Jr., eds., *Proceedings of the Second Annual Meeting of the Berkeley Linguistics Society*, 202–211. Berkeley: Berkeley Linguistics Society, University of California.

Heim, Irene. 2000. Degree operators and scope. In Brendan Jackson and Tanya Matthews, eds., *Proceedings of Semantics and Linguistic Theory (SALT) 10*, 40–64. Ithaca, NY: CLC Publications, Department of Linguistics, Cornell University.

Heim, Irene. 2006. Little. In Christopher Tancredi, Makoto Kanazawa, Ikumi Imani, and Kiyomi Kusumoto, eds., *Proceedings of Semantics and Linguistic Theory (SALT) 16*, 35–58. Ithaca, NY: CLC Publications, Department of Linguistics, Cornell University.

Helke, Kyle. 2011. Arguments and agreement in Tewa. Ms., Queen Mary University of London.

Hernanz, Ma. Lluïsa. 1999. El infinitivo. In Ignacio Bosque and Violeta Demonte, eds., *Gramática descriptiva de la lengua española*, 2197–2356. Madrid: Espasa.

Hewitt, George. 1979. *Lingua Descriptive Studies 2: Abkhaz*. Amsterdam: North-Holland.

Hilpert, Martin. 2008. The English comparative—language structure and language use. *English Language and Linguistics* 12:395–417.

Hiraiwa, Ken. 2001. Multiple agree and the defective intervention constraint in Japanese. In Ora Matushansky, ed., *MITWPL 40: Harvard-MIT Student Conference on Language Research (HUMIT) 1*, 67–80. Cambridge, MA: MITWPL, Department of Linguistics and Philosophy, MIT.

Hockett, Charles F. 1954. Two models of grammatical description. *Word* 10:210–131.

Hoeksema, Jack. 2012. Elative compounds in Dutch. In Guido Oebel, ed., *Cross-Linguistic Comparison of Intensified Adjectives and Adverbs*, vol. 2, 97–142. Hamburg: Verlag Dr. Kovac.

Hualde, José Ignacio. 2003. Finite forms. In José Ignacio Hualde and Jon Ortiz de Urbina, eds., *A Grammar of Basque*, 205–246. Berlin: Mouton de Gruyter.

Hutchisson, Don. 1986. Sursurunga pronouns and the special uses of quadral number. In Ursula Wiesemann, ed., *Pronominal Systems*, 217–255. Tübingen: Narr.

Hyman, Larry, Sharon Inkelas, and Galen Sibanda. 2008. Morphosyntactic correspondence in Bantu reduplication. In Kristin Hanson and Sharon Inkelas, eds., *The Nature of the Word: Essays in Honor of Paul Kiparsky*, 273–309. Cambridge, MA: MIT Press.

Iatridou, Sabine, Elena Anagnostopoulou, and Roumyana Izvorski. 2001. Observations about the form and meaning of the perfect. In Michael Kenstowicz, ed., *Ken Hale: A Life in Language*, 189–238. Cambridge, MA: MIT Press. Reprinted in A. Alexiadou, M. Rathert, and A. von Stechow, eds., *Perfect Explorations*, Berlin: Mouton de Gruyter, 2003.

Idsardi, William. 1992. The computation of prosody. Doctoral dissertation, MIT.

Inkelas, Sharon, and Cheryl Zoll. 2007. Is grammar dependence real? A comparison between cophonological and indexed constraint approaches. *Linguistics* 45:133–171.

Ippolito, Michela M. 1999. On the past participle morphology in Italian. In Karlos Arregi, Vivian Lin, Cornelia Krause, and Benjamin Bruening, eds., *MITWPL 33: Papers in Morphology and Syntax, Cycle One*, 111–137. Cambridge, MA: MITWPL, Department of Linguistics and Philosophy, MIT.

Itô, Junko, and Armin Mester. 1999. The phonological lexicon. In N. Tsujimura, ed., *The Handbook of Japanese Linguistics*, 62–100. Malden, MA: Blackwell.

Jackendoff, Ray. 1977. *X-Bar Syntax: A Study of Phrase Structure*. Cambridge, MA: MIT Press.

Jackendoff, Ray. 2000. Curiouser and curiouser. *Snippets* 1:8.

Jacobs, Bart. 2011. Present and historical perspectives on the Catalan *go*-past. *Zeitschrift für Katalanistik* 24:227–255.

Janda, Laura A., and Charles Townsend. 2000. *Czech*. Berlin: Lincom Europa.

Jespersen, Otto. 1956. *A Modern English Grammar on Historical Principles*. 7th ed. Copenhagen: Ejnar Munksgaard. Originally published in 1909.

Johannessen, Janne Bondi. 1993. Coordination: A minimalist approach. Doctoral dissertation, University of Oslo.

Johnson, Kyle. 2000. How far will quantifiers go? In Roger Martin, David Michaels, and Juan Uriagereka, eds., *Step by Step: Essays on Minimalist Syntax in Honor of Howard Lasnik*, 187–210. Cambridge, MA: MIT Press.

Jovanović, Vladimir Ž. 2009. Certain morpho-semantic implications with the grammatical category of comparison in English. *Facta Universitatis* 7:19–28.

Juge, Matthew. 2006. Morphological factors in the grammaticalization of the Catalan 'go'-past. *Diachronica* 23(2):313–339.

Kagan, Olga, and Sascha Alexeyenko. 2011. Degree modification in Russian morphology: The case of the suffix *-ovat*. In Ingo Reich, Eva Horch, and Dennis Pauly, eds., *Proceedings of Sinn und Bedeutung 15*, 321–335. Saarbrücken: Saarland University Press.

Kager, René. 1995. English stress: Re-inventing the paradigm (Review of *Principles of English Stress* by Luigi Burzio). *GLOT International* 9/10:19–21.

Kautzsch, Emil. 1910. *Genesius' Hebrew Grammar*. 2nd ed. Oxford: Clarendon Press.

Kayne, Richard. 1981. Two notes on the NIC. In Adriana Belletti and Luigi Rizzi, eds., *Theory of Markedness in Generative Grammar*, 317–346. Pisa: Scuole Normale Superiore.

Kayne, Richard. 1998. Overt vs. covert movement. *Syntax* 1:128–191.

Kennedy, Christopher. 1997. Antecedent-contained deletion and the syntax of quantification. *Linguistic Inquiry* 28:662–688.

Kennedy, Christopher. 1999. *Projecting the Adjective: The Syntax and Semantics of Gradability and Comparison*. New York: Garland.

Kennedy, Christopher, and Peter Svenonius. 2006. Northern Norwegian degree questions and the syntax of measurement. In Mara Frascarelli, ed., *Phases of Interpretation*, 129–157. Berlin: Mouton de Gruyter.

Kharytonava, Olga. 2011. *Noms composées en turc et morpheme -(s)I*. Doctoral dissertation, University of Western Ontario.

Kiefer, Ferenc. 1978. Adjectives and presuppositions. *Theoretical Linguistics* 5:135–173.

Kiefer, Ferenc. 2001. Morphology and pragmatics. In Andrew Spencer and Arnold M. Zwicky, eds., *The Handbook of Morphology*, 272–279. Oxford: Blackwell.

Kikuchi, Seiichiro. 2001. Spanish definite article allomorphy: A correspondence approach. In Phonological Society of Japan, ed., *Phonological Studies 4*, 49–56. Tokyo: Kaitakusha.

Kiparsky, Paul. 1973. "Elsewhere" in phonology. In Stephen R. Anderson and Paul Kiparsky, eds., *Festschrift for Morris Halle*, New York: Holt, Rinehart and Winston.

Kiparsky, Paul. 1979. Metrical structure assignment is cyclic. *Linguistic Inquiry* 10:421–441.

Kiparsky, Paul. 1982. From cyclic phonology to lexical phonology. In H. van der Hulst and N. Smith, eds., *The Structure of Phonological Representations*, 130–175. Dordrecht: Foris.

Kiparsky, Paul. 1996. Allomorphy or morphophonology? In Rajendra Singh and Richard Desrochers, eds., *Trubetzkoy's Orphan*, 13–31. Amsterdam: John Benjamins.

Kiparsky, Paul. 2005. Blocking and periphrasis in inflectional paradigms. In *Yearbook of Morphology 2004*, 113–135. Dordrecht: Springer.

van Koppen, Marjo. 2005. One probe—two goals: Aspects of agreement in Dutch dialects. Doctoral dissertation, Universiteit Leiden.

Krasikova, Sveta. 2009. Norm-relatedness in degree constructions. In Arndt Riester and Torgrim Solstad, eds., *Proceedings of Sinn und Bedeutung 13*, 293–308. Stuttgart: University of Stuttgart.

Krasikova, Sveta. 2010. Scales in the meaning of adjectives. *Oslo Studies in Language* 2:51–74.

Kratzer, Angelika. 1996. Severing the external argument from its verb. In Johan Rooryck and Laurie Zaring, eds., *Phrase Structure and the Lexicon*, 109–137. Dordrecht: Kluwer.

Kratzer, Angelika. 1998. More structural analogies between pronouns and tenses. In D. Strolovitch and A. Lawson, eds., *Proceedings from Semantics and Linguistic Theory (SALT) VIII*, 92–110. Ithaca, NY: CLC Publications, Department of Linguistics, Cornell University.

Krifka, Manfred. 1992. Thematic relations as links between nominal reference and temporal constitution. In Ivan Sag and Anna Szabolcsi, eds., *Lexical Matters*, 29–53. Stanford, CA: CSLI.

Kroch, Anthony. 1994. Morphosyntactic variation. In Katharine Beals, Jeannette Denton, Robert Knippen, Lynette Melnar, Hisami Suzuki, and Erica Zeinfeld, eds., *Papers from the 30th Regional Meeting of the Chicago Linguistic Society: Parasession on Variation and Linguistic Theory*, 180–201. Chicago: Chicago Linguistic Society.

Kroskrity, Paul V. 1985. A holistic understanding of Arizona Tewa passives. *Language* 61:306–328.

Kroskrity, Paul V. 2010. The art of voice: Understanding the Arizona Tewa inverse in its grammatical, narrative, and language-ideological contexts. *Anthropological Linguistics* 52:49–79.

Kroskrity, Paul V., and Dewey Healing. 1978. Coyote and bullsnake. In William Bright, ed., *Coyote Stories*, 162–171, Chicago: University fo Chicago Press.

Kytö, Merja, and Suzanne Romaine. 1997. Competing forms of adjective comparison in modern English: What could be more quicker and easier and more effective? In Terttu Nevalainen and Leena Kahlas-Tarkka, eds., *To Explain the Present: Studies in the Changing English Language in Honour of Mattie Rissanen*, 329–352. Helsinki: Société Néophilologique.

Labov, William. 1969. Contraction, deletion, and inherent variability of the English copula. *Language* 45:715–762.

Labov, William. 1972. *Sociolinguistic Patterns*. Philadelphia: University of Pennsylvania Press.

Laka, Itziar. 1993. The structure of inflection: A case study in X° syntax. In José Ignacio Hualde and Jon Ortiz de Urbina, eds., *Generative Studies in Basque Linguistics*, 21–70. Amsterdam: John Benjamins.

Lebeaux, David. 1995. Where does binding theory apply? In Ricardo Echepare and Viola Miglio, eds., *Papers in Syntax, Syntax-Semantics Interface and Phonology*, 63–88. College Park: University of Maryland Working Papers in Linguistics.

Legate, Julie Anne. 2003. Some interface properties of the phase. *Linguistic Inquiry* 34:506–516.

Levinson, Lisa. 2010. Arguments for pseudo-resultative predicates. *Natural Language & Linguistic Theory* 28:135–182.

Liberman, Mark. 1975. The intonational system of English. Doctoral dissertation, MIT.

Liberman, Mark, and Alan Prince. 1977. On stress and linguistic rhythm. *Linguistic Inquiry* 8:249–336.

Lieber, Rochelle. 1980. The organization of the lexicon. Doctoral dissertation, MIT.

Lieber, Rochelle. 1987. *An Integrated Theory of Autosegmental Processes*. Albany: State University of New York Press.

Lieber, Rochelle. 1992. *Deconstructing Morphology: Word Formation in Syntactic Theory*. Chicago: University of Chicago Press.

Lieber, Rochelle. 2010. *Introducing Morphology*. Cambridge: Cambridge University Press.

Lindquist, Hans. 2000. Livelier or more lively? Syntactic and contextual factors influencing the comparison of disyllabic adjectives. In John M. Kirk, ed., *Corpora Galore: Analyses and Techniques in Describing English, Language and Computers: Studies in Practical Linguistics 30*, 125–132. Amsterdam: Rodopi.

Link, Godehard. 1983. The logical analysis of plurals and mass terms: A lattice-theoretical approach. In Rainer Bäuerle, Christoph Schwarze, and Arnim von Stechow, eds., *Meaning, Use and Interpretation of Language*, 302–323. Berlin: Walter de Gruyter.

Lomashvili, Leila, and Heidi Harley. 2011. Phases and templates in Georgian agreement. *Studia Linguistica* 65, 233–267.

Longobardi, Giuseppe. 1992. In defense of the correspondence hypothesis: Island effects and parasitic constructions in logical form. In C. T. James Huang and Robert May, eds., *Logical Structure and Linguistic Structure: Cross-Linguistic Perspectives*, 149–196. Dordrecht: Kluwer Academic Publishers.

Lowenstamm, Jean. 2010. Derivational affixes as roots (phasal spellout meets English stress shift). Ms., Université Paris-Diderot and CNRS.

Lynch, John, Malcolm Ross, and Terry Crowley. 2002. Typological overview. In John Lynch, Malcolm Ross, and Terry Crowley, eds., *The Oceanic Languages*, 34–53. Surrey: Curzon Press.

Mace, John. 2003. *Persian Grammar*. London: Routledge Curzon.

Maiden, Martin. 1991. *Interactive Morphology: Metaphony in Italy*. London: Routledge.

Maiden, Martin. 1992. Irregularity as a determinant of morphological change. *Journal of Linguistics* 28:285–312.

Maiden, Martin. 2004a. Morphological autonomy and diachrony. *Yearbook of Morphology* 2004:137–175.

Maiden, Martin. 2004b. When lexemes become allomorphs—On the genesis of suppletion. *Folia Linguistica* 38 (3–4):227–256.

Maiden, Martin, John Charles Smith, Maria Goldbach, and Marc-Olivier Hinzelin, ed. 2011. *Morphological Autonomy: Perspectives from Romance Inflectional Morphology*. Oxford: Oxford University Press.

Marantz, Alec. 1988. Clitics, morphological merger, and the mapping to phonological structure. In Michael Hammond and Michael Noonan, eds., *Theoretical Morphology*, 253–270. San Diego: Academic Press.

Marantz, Alec. 1989. Relations and configurations in Georgian. Ms., University of North Carolina, Chapel Hill.

Marantz, Alec. 1991. Case and licensing. In German F. Westphal, Benjamin Ao, and Hee-Rahk Chae, eds., *Proceedings of the 8th Eastern States Conference on Linguistics*, 234–253. Columbus: Ohio State University.

Marantz, Alec. 1992. What kind of pieces are inflectional morphemes? Paper presented at the Berkeley Linguistics Society, February.

Marantz, Alec. 1994. A late note on late insertion. In Young-Suk Kim, Byung-Choon Lee, Kyoung-Jae Lee, Hyun-Kwon Yang, and Jong-Yurl Yoon, eds., *Explorations in Generative Grammar: A Festschrift for Dong-Whee Yang*, 396–413. Seoul: Hankuk.

Marantz, Alec. 1995. "Cat" as a phrasal idiom. Ms., MIT.

Marantz, Alec. 1997. No escape from syntax: Don't try morphological analysis in the privacy of your own lexicon. In A. Dimitriadis, L. Siegel, C. Surek-Clark, and A. Williams, eds., *Proceedings of the 21st Annual Penn Linguistics Colloquium*, 201–225. Philadelphia: Penn Linguistics Club.

Marantz, Alec. 1999. The pieces of derivation. Handout, MIT.

Marantz, Alec. 2001. Words and things. Handout, MIT.

Marantz, Alec. 2007. Phases and words. In Sook-hee Choe, ed., *Phases in the Theory of Grammar*, Seoul: Dong In Publisher.

Marantz, Alec. Forthcoming. Argument structure. *Lingua*.

Martin, Roger Andrew. 1996. A minimalist theory of PRO and control. Doctoral dissertation, University of Connecticut, Storrs.

Martin, Roger Andrew. 2001. Null case and the distribution of PRO. *Linguistic Inquiry* 32:141–166.

Martinez, Esther, ed. 1983. *San Juan Pueblo Téwa Dictionary*. Portales, NM: Bishop Publishing Co. A production of the San Juan Pueblo Bilingual Program.

Marušič, Franc Lanko. 2009. Non-simultaneous spell-out in the clausal and nominal domain. In Kleanthes K. Grohmann, ed., *InterPhases: Phase-theoretic Investigations of the Linguistic Interfaces*, 151–181. Oxford: Oxford University Press.

Marušič, Franc Lanko, Andrew Nevins, and William Badecker. Forthcoming. The grammars of conjunction agreement in Slovenian. *Syntax.*

Marvà, Jeroni. 1968. *Curs superior de gramàtica catalana.* Barcelona: Barcino.

Marvin, Tatjana. 2002. Topics in the stress and syntax of words. Doctoral Dissertation, MIT.

Mascaró, Joan. 1986. *Morfologia.* Barcelona: Enciclopèdia Catalana.

Mascaró, Joan. 2002a. El sistema vocàlic: Reducció vocàlica. In Joan Solà, Maria Rosa Lloret, Joan Mascaró, and Manuel Pérez Saldanya, eds., *Gramàtica del català contemporani*, 89–123. Barcelona: Empúries.

Mascaró, Joan. 2002b. Morfologia: Aspectes generals. In Joan Solà, Maria Rosa Lloret, Joan Mascaró, and Manuel Pérez Saldanya, eds., *Gramàtica del català contemporani*, 465–482. Barcelona: Empúries.

Mascaró, Joan. 2011. The realization of features in asymmetric agreement in DPs. *Snippets* 23:11–12.

Matthews, Peter H. 1965. The inflectional component of a word-and-paradigm grammar. *Journal of Linguistics* 1:139–171.

Matthews, Peter H. 1972. *Inflectional Morphology: A Theoretical Study Based on Aspects of Latin Verb Conjugation.* Cambridge: Cambridge University Press.

Matushansky, Ora. 2001. The more the merrier: The syntax of synthetic and analytic comparatives. Paper presented at GLOW 24, Braga, Portugal.

Matushansky, Ora. 2002. Movement of degree/degree of movement. Doctoral dissertation, MIT. Cambridge, MA: MITWPL, Department of Linguistics and Philosophy, MIT.

Matushansky, Ora. 2008. On the attributive nature of superlatives. *Syntax* 11:26–90.

McCarthy, John J. 1981. A prosodic theory of nonconcatenative morphology. *Linguistic Inquiry* 12:373–418.

McCarthy John, and Alan Prince. 1993. Prosodic morphology, I: Constraint interaction and satisfaction. Ms. University of Massachusetts, Amherst, and Rutgers University, New Brunswick.

McGinnis, Martha. 1996. Two kinds of blocking. In Hee-don Ahn, Myung-Yoon Kang, Yong-Suck Kim, and Sookhee Lee, eds., *Morphosyntax in Generative Grammar: Proceedings of the 1996 Seoul International Conference on Generative Grammar*, 357–368. Seoul: Hankuk.

McGinnis, Martha. 1998. Locality in A-movement. Doctoral dissertation, MIT.

McGinnis, Martha. 2004. Lethal ambiguity. *Linguistic Inquiry* 35:47–95.

McGinnis, Martha. 2005. On markedness asymmetries in person and number. *Language* 81:699–718.

McGinnis, Martha. 2008. Phi-feature competition in morphology and syntax. In Daniel Harbour, David Adger, and Susana Béjar, eds., *Phi-Theory: Phi-Features across*

Modules and Interfaces, 155–184. Oxford Studies in Theoretical Linguistics. Oxford: Oxford University Press.

Mithun, Marianne. 1988. Lexical categories and the evolution of number marking. In Michael Hammond and Michael Noonan, eds., *Theoretical Morphology: Approaches in Modern Linguistics*, 211–234. San Diego: Academic Press.

Miyagawa, Shigeru. 1998. (S)ase as an elsewhere causative and the syntactic nature of words. *Journal of Japanese Linguistics* 16:67–110.

Miyagawa, Shigeru. 1999. Causatives. In Natsuko Tsujimura, ed., *The Handbook of Japanese Linguistics*, 236–268. Oxford: Blackwell.

Moll, Francesc de B. 1968. *Gramàtica catalana (referida especialment a les Illes Balears)*. Palma de Mallorca: Editorial Moll.

Mondorf, Britta. 2002. The effect of prepositional complements on the choice of synthetic or analytic comparatives. In Hubert Cuyckens and Günter Radden, eds., *Perspectives on Prepositions*, 65–78. Tübingen: Niemeyer.

Mondorf, Britta. 2003. Support for *more*-support. In Günter Rohdenburg and Britta Mondorf, eds., *Determinants of Grammatical Variation in English*, 251–304. Topics in English Linguistics 43. Berlin: Mouton de Gruyter.

Mondorf, Britta. 2007. Recalcitrant problems of comparative alternation and new insights emerging from Internet data. In Marianne Hundt, Nadja Nesselhauf, and Carolin Biewer, eds., *Corpus Linguistics and the Web*, 211–232. Amsterdam: Rodopi.

Mondorf, Britta. 2009. More Support for *More*-Support: The Role of Processing Constraints on the Choice between Synthetic and Analytic Comparative Forms. *Studies in Language Variation* 4. Amsterdam: John Benjamins.

Munn, Alan. 1992. A null operator analysis of ATB gaps. *Linguistic Review* 9:1–26.

Nash, Léa. 1994. On *be* and *have* in Georgian. In Heidi Harley and Colin Phillips, eds., *MITWPL 22: The Morphology-Syntax Connection*, 153–171. Cambridge, MA: MITWPL, Department of Linguistics and Philosophy, MIT.

Nash, Léa. 1995. Portée argumentale et marquage casuel dans les langues SOV et dans les langues ergatives: L'example du géorgien. Doctoral dissertation, Université de Paris 8.

Nevins, Andrew. 2002a. Fission without a license. Talk presented at Concordia University, Montreal, November 15.

Nevins, Andrew. 2002b. When "we" disagrees in circumfixes. Handout from a talk presented at LASSO 31, Los Angeles, October 4–6.

Nevins, Andrew. 2007. The representation of third person and its consequences for person-case effects. *Natural Language & Linguistic Theory* 25:273–313.

Nevins, Andrew. 2011a. Convergent evidence for rolling up Catalan adjectives. *Linguistic Inquiry* 42:339–345.

Nevins, Andrew. 2011b. Marked targets versus marked triggers and impoverishment of the dual. *Linguistic Inquiry* 42:413–444.

Nevins, Andrew. 2011c. Phonologically conditioned allomorph selection. In Marc van Oostendorp, Colin J. Ewen, Elizabeth Hume, and Keren Rice, eds., *Blackwell Companion to Phonology*, 2357–2382. Malden, MA: Wiley-Blackwell.

Nevins, Andrew, and Jeffrey K. Parrott. 2010. Variable rules meet Impoverishment theory: Patterns of agreement leveling in English varieties. *Lingua* 120:1135–1159.

Nevins, Andrew, and Filomena Sandalo. 2011. Markedness and morphotactics in Kadiwéu [+participant] agreement. *Morphology* 21:351–378.

Newell, Heather. 2005. Bracketing paradoxes and particle verbs: A late adjunction analysis. In Sylvia Blaho, Luis Vicente, and Erik Schoorlemmer, eds., *Proceedings of ConSOLE XIII (2004, Tromsø)*, 249–272. Leiden: LUCL.Noyer.

Noyer, Rolf. 1992. Features, positions, and affixes in autonomous morphological structure. Doctoral dissertation, MIT. Revised version published by Garland, New York, 1997.

Ogihara, Toshiyuki. 1995. The semantics of tense in embedded clauses. *Linguistic Inquiry* 26:663–679.

Ogihara, Toshiyuki. 1996. *Tense, Attitude, and Scope*. Dordrecht: Kluwer Academic Publishers.

Ogihara, Toshiyuki. 2007. Tense and aspect in truth-conditional semantics. *Lingua* 117:392–418.

Oltra-Massuet, Isabel. 1999. On the constituent structure of Catalan verbs. In Karlos Arregi, Vivian Lin, Cornelia Krause, and Benjamin Bruening, eds., *MITWPL 33: Papers in Morphology and Syntax, Cycle One*, 279–322. Cambridge, MA: MITWPL, Department of Linguistics and Philosophy, MIT.

Oltra-Massuet, Isabel. 2000. On the notion of theme vowel: A new approach to Catalan verbal morphology. *MIT Occasional Papers in Linguistics* 19. Cambridge, MA: MITWPL, Department of Lingustics and Philosophy, MIT.

Oltra-Massuet, Isabel, and Karlos Arregi. 2005. Stress-by-structure in Spanish. *Linguistic Inquiry* 36:43–84.

Orgun, Cemil Orhan. 1996. Sign-based morphology and phonology: With special attention to Optimality Theory. Doctoral dissertation, University of California, Berkeley.

Ormazabal, Javier, and Juan Romero. 2007. The object agreement constraint. *Natural Language & Linguistic Theory* 25:315–347.

Pater, Joe. 2000. Non-uniformity in English secondary stress: The role of ranked and lexically specific constraints. *Phonology* 17:237–274.

Pater, Joe. 2010. Morpheme-specific phonology: Constraint indexation and inconsistency resolution. In Steve Parker, ed., *Phonological Argumentation: Essays on Evidence and Motivation*, 123–154. London: Equinox.

Perea, M. Pilar. 1999. *Compleció i ordenació de "La flexió verbal en els dialectes catalans" d'A. M. Alcover i F. de B. Moll*. Barcelona: Institut d'Estudis Catalans. CD edition.

Perea, M. Pilar. 2002. Flexió verbal regular. In Joan Solà, Maria Rosa Lloret, Joan Mascaró, and Manuel Pérez Saldanya, eds., *Gramàtica del català contemporani*, 583–646. Barcelona: Empúries.

Pérez Saldanya, Manuel. 1996. Gramaticalització i reanàlisi: El cas del pretèrit perfet perifràstic en català. In Axel Schönberger and Tilbert Dídac Stegmann, eds., *Actes del*

Desè Col·loqui Internacional de Llengua i Literatura Catalanes, vol. 3, 71–107. Barcelona: Publicacions de l'Abadia de Montserrat.

Pérez Saldanya, Manuel. 1998. *Del llatí al català: Morfosintaxi verbal històrica.* València: Universitat de València.

Pérez Saldanya, Manuel. 2002. Les relacions temporals i aspectuals. In Joan Solà, Maria Rosa Lloret, Joan Mascaró, and Manuel Pérez Saldanya, eds., *Gramàtica del català contemporani,* 2567–2662. Barcelona: Empúries.

Perlmutter, David. 1971. *Deep and Surface Structure Constraints in Syntax.* New York: Holt, Rinehart and Winston.

Pesetsky, David. 1979. Russian morphology and lexical theory. Ms., MIT. http://web. mit.edu/linguistics/www/pesetsky/russmorph.pdf.

Pesetsky, David. 1985. Morphology and logical form. *Linguistic Inquiry* 16:193–246.

Pesetsky, David. 1992. Zero syntax II: An essay on infinitives. Ms., MIT.

Pesetsky, David, and Esther Torrego. 2007. The syntax of valuation and the interpretability of features. In Simin Karimi, Vida Samiian, and Wendy Wilkins, eds., *Phrasal and Clausal Architecture,* 262–294. Amsterdam: John Benjamins.

Picallo, M. Carme. 2008. Gender and number in Romance. *Lingue & Linguaggio* 7:47–61.

Pinker, Steven. 1999. *Words and Rules: The Ingredients of Language.* New York: Perseus Books.

Pinker, Steven, and Michael Ullman. 2002. The past and future of the past tense. *Trends in Cognitive Sciences* 6:456–463.

Plag, Ingo. 1998. Morphological haplology in a constraint-based morpho-phonology. In Wolfgang Kehrein and Richard Wiese, eds., *Phonology and Morphology of the Germanic Languages,* 199–215. Tübingen: Niemeyer.

Plag, Ingo. 1999. *Morphological Productivity.* Berlin: Mouton de Gruyter.

Poletto, Cecilia. 2000. *The Higher Functional Field.* Oxford: Oxford University Press.

Pomino, Natascha, and Eva-Maria Remberger. 2008. Distributed Morphology and minimalist syntax of Spanish *ir.* Ms., Universität Konstanz.

Poser, William J. 1992. Blocking of phrasal constructions by lexical items. In Ivan A. Sag and Anna Szabolcsi, eds., *Lexical Matters,* 111–130. Stanford, CA: CSLI.

Prince, Alan. 1983. Relating to the grid. *Linguistic Inquiry* 14:19–100.

Prince, Alan, and Paul Smolensky. [1993] 2004. *Optimality Theory: Constraint Interaction in Generative Grammar.* Malden, MA: Blackwell.

Pullum, Geoffrey K., and Arnold M. Zwicky. 1991. A misconceived approach to morphology. In Dawn Bates, ed., *Proceedings of the West Coast Conference on Formal Linguistics,* 10, 387–398. Stanford, CA: CSLI, WCCFL.

Pylkkänen, Liina. 2008. *Introducing Arguments.* Cambridge, MA: MIT Press.

Quirk, Randolph, Sidney Greenbaum, Geoffrey Leech, and Jan Svartvik. 1985. *A Comprehensive Grammar of the English Language.* New York: Longman.

Radkevich, Nina. 2010. On location: The structure of case and adpositions. Doctoral dissertation, thesis, University of Connecticut, Storrs.

Rasom, Sabrina. 2008. Lazy concord in the Central Ladin feminine DP: A case study on the interaction between morphosyntax and semantics. Doctoral dissertation, Università degli Studi di Padova.

Redden, James. 1966. Walapai II: Morphology. *International Journal of American Linguistics* 23:141–163.

Rett, Jessica. 2008. Degree modification in natural language. Doctoral dissertation, Rutgers University.

Rezac, Milan. 2003. The fine structure of cyclic agree. *Syntax* 6:156–182.

Rezac, Milan. 2008a. The forms of dative displacement: From Basauri to Itelmen. In Xabier Artiagoitia and Joseba Andoni Lakarra, eds., *Gramatika Jaietan: Festschrift for Patxi Goenaga*, 709–724. Bilbo: Euskal Herriko Unibertsitatea.

Rezac, Milan. 2008b. Phi-Agree and theta-related case. In Daniel Harbour, David Adger, and Susana Béjar, eds., *Phi Theory: Phi-Features across Modules and Interfaces*, 83–129. Oxford: Oxford University Press.

Ringe, Don. 2006. *From Proto-Indo-European to Proto-Germanic*. Oxford: Oxford University Press.

Ritter, Elizabeth. 1997. Agreement in the Arabic prefix conjugation: Evidence for a non-linear approach to person, number and gender features. In Leslie Blair, Christine Burns, and Lorna Rowsell, eds., *Proceedings of the Canadian Linguistic Association Conference*, 191–202. Calgary: Calgary Working Papers in Linguistics, University of Calgary.

Rix, Helmut, Martin Kümmel, Thomas Zehnder, Reiner Lipp, and Brigitte Schirmer. 2001. *LIV: Lexikon der indogermanischen Verben: die Wurzeln und ihre Primärstammbildungen*. 2nd. Wiesbaden: Reichert.

Rose, Sharon. 2007. Chaha (Gurage) morphology. In Alan S. Kaye, ed., *Morphologies of Africa and Asia*, vol. 1, 403–428. Winona Lake, IN: Eisenbrauns.

Ross, John. 1967. Constraints on variables in syntax. Doctoral dissertation, MIT.

Ross, Malcolm. 2002. Mussau. In John Lynch, Malcolm Ross, and Terry Crowley, eds., *The Oceanic Languages*, 148–166. Surrey: Curzon Press.

Sadock, Jerrold M. 1991. *Autolexical Syntax: A Theory of Parallel Grammatical Representations*. Chicago: University of Chicago Press.

Salanova, Andrés Pablo. 2004. Subtractive morphology in Mêbengokre. Ms., MIT.

Samek-Lodovici, Vieri. 2002. Agreement Impoverishment under Subject Inversion: A crosslinguistic analysis. In Gisbert Fanselow and Caroline Féry, eds., *Resolving Conflicts in Grammar. Linguistische Berichte* Sonderhaft 11:49–82.

Sanchís Guarner, Manuel. 1950. *Gramàtica valenciana*. València: Torre. Cited from the facsimile edited by Antoni Ferrando. Barcelona: Alta Fulla, 1993.

Sauerland, Uli. 2003. Intermediate adjunction with A-movement. *Linguistic Inquiry* 34:308–314.

Scatton, Ernest. 1984. *A Reference Grammar of Modern Bulgarian*. Columbus, OH: Slavica Publishers.

Schäfer, Florian. 2008. *The Syntax of (Anti-)Causatives: External Arguments in Change-of-State Contexts*. Amsterdam: John Benjamins.

Scheer, Tobias. 2008. Spell out your sister! In Natasha Abner and Jason Bishop, eds., *Proceedings of the 27th West Coast Conference on Formal Linguistics (WCCFL)*, 379–387. Somerville, MA: Cascadilla Proceedings Project.

Schindler, Jochem. 1972. L'Apophonie des noms-racine indo-européens. *Bulletin de la Sociéte de Linguistique* 67:31–38.

Schindler, Jochem. 1975a. L'Apophonie des thèmes indo-européens en -r/-n. *Bulletin de la Sociéte de Linguistique* 70:1–10.

Schindler, Jochem. 1975b. Zum ablaut der neutralen *s*-Stämme des Indogermanischen. In Helmut Rix, ed., *Flexion und Wortbildung*, 259–267. Wiesbaden: Reichert.

Schoorlemmer, Erik. 2009. Agreement, dominance and doubling: The morphosyntax of DP. Doctoral dissertation, University of Leiden.

Schwarzschild, Roger, and Karina Wilkinson. 2002. Quantifiers in comparatives: A semantics of degree based on intervals. *Natural Language Semantics* 10:1–41.

Schweizer, Bruno. 2008. *Zimbrische Gesamtgrammatik*. Stuttgart: Franz Steiner Verlag.

Sedighi, Anousha. 2005. Animacy: The overlooked feature in Persian. In Marie-Odile Junker, Martha McGinnis, and Yves Roberge, eds., *Proceedings of the 2004 Annual Conference of the Canadian Linguistic Association*. http://homes.chass.utoronto.ca/~cla-acl/actes2004/actes2004.html.

Seidenberg, Mark S., and Laura M. Gonnerman. 2000. Explaining derivational morphology as the convergence of codes. *Trends in Cognitive Sciences* 4:353–361.

Seuren, Pieter A. M. 1984. The comparative revisited. *Journal of Semantics* 3:109–141.

Sharvit, Yael, and Penka Stateva. 2002. Superlative expressions, context, and focus. *Linguistics and Philosophy* 25:453–505.

Shlonsky, Ur. 2004. The form of Semitic noun phrases. *Lingua* 114 (2): 1465–1526.

Siddiqi, Daniel. 2009. *Syntax within the Word: Economy, Allomorphy, and Argument Selection in Distributed Morphology*. Amsterdam: John Benjamins.

Siegel, Dorothy. 1974. Topics in English morphology. Doctoral dissertation, MIT.

Siewierska, Anna. 2004. *Person*. Cambridge: Cambridge University Press.

Simon, Dylan A., Gwyneth Lewis, and Alec Marantz. 2012. Disambiguating form and lexical frequency effects in MEG responses using homonyms. *Language and Cognitive Processes* 27:275–287.

Sproat, Richard. 1985. On deriving the lexicon. Doctoral dissertation, MIT.

Stateva, Penka. 2000a. In defense of the movement theory of superlatives. In Rebecca Daly and Anastasia Riehl, eds., *Proceedings of the Eastern States Conference on Linguistics 1999*, 215–226. Ithaca, NY: CLC Publications.

Stateva, Penka. 2000b. Towards a superior theory of superlatives. In Christine Czinglar, Katharina Köhler, Erica Thrift, Erik Jan van der Torre, and Malte Zimmermann, eds., *ConSole VIII Proceedings: Proceedings of the Eighth Conference of the Student Organization of Linguistics in Europe*, 313–326. Leiden: SOLE.

Stateva, Penka. 2002. How different are different degree constructions? Doctoral dissertation, University of Connecticut, Storrs.

Stateva, Penka. 2003. Superlative *more*. In Robert B. Young and Yuping Zhou, eds., *Proceedings of Semantics and Linguistic Theory (SALT) 13*, 276–291. Ithaca, NY: CLC Publications, Department of Linguistics, Cornell University.

Steriade, Donca. 1999. Lexical conservatism in French adjectival liaison. In B. Bullock, M. Authier, and L. Reed, eds., *Formal Perspectives in Romance Linguistics*, 243–270. Amsterdam: John Benjamins.

Stockall, Linnea, and Alec Marantz. 2006. A single route, full decomposition model of morphological complexity: MEG evidence. *Mental Lexicon* 7 (1):85–123.

Stump, Gregory. 2001. *Inflectional Morphology: A Theory of Paradigm Structure*. Cambridge: Cambridge University Press.

Svenonius, Peter. 2008. The position of adjectives and other phrasal modifiers in the decomposition of DP. In Louise McNally and Chris Kennedy, eds., *Adjectives and Adverbs: Syntax, Semantics, and Discourse*, 16–42. Oxford: Oxford University Press.

Svenonius, Peter. 2012a. Look both ways: Outward-looking allomorphy in Icelandic participles. lingbuzz/001519.

Svenonius, Peter. 2012b. Spanning. Ms., University of Tromsø. lingBuzz/001501, http://ling.auf.net/lingbuzz.

De Swart, Henriëtte. 1998. Aspect shift and coercion. *Natural Language & Linguistic Theory* 16:347–38.

Takahashi, Masahiko. 2010. Case, phases, and nominative/accusative conversion in Japanese. *Journal of East Asian Linguistics* 19:319–355.

Tewa Tales. 1982. *Téwa Pehtsiye: Tewa Tales, San Juan Dialect*. A production of the San Juan Pueblo Bilingual Program.

Todolí, Júlia. 1992. Variants dels pronoms febles de 3a persona al País Valencià: Regles fonosintàctiques i morfológiques subjacents. *Zeitschrift für Katalanistik* 5:137–160.

Tourabi, Abderrezzak. 2002. Arabic subject-verb agreement affixes: Morphology, specification and Spell-Out. In Aniko Csirmaz, Zhiqiang Li, Andrew Nevins, Olga Vaysman, and Michael Wagner, eds., *MITWPL 42: Phonological Answers*. Cambridge, MA: MITWPL, Department of Linguistics and Philosophy, MIT.

Trommer, Jochen. 2008. Direction marking and case in Menominee. In Daniel Harbour, David Adger, and Susana Béjar, eds., *Phi Theory: Phi Features across Modules and Interfaces*, 221–250. Oxford: Oxford University Press.

University of Latvia. 1999. *Latvian Grammar* (online). http://www.ailab.lv/valoda/v1.htm.

Vallduví, Enric. 1988. Sobre el perfet perifràstic del català. In Philip D. Rasico and Curt Wittlin, eds., *Actes del Cinquè Col·loqui d'Estudis Catalans a Nord-Amèrica*, 85–98. Barcelona: Publicacions de l'Abadia de Montserrat.

Viaplana, Joaquim, Maria-Rosa Lloret, Maria-Pilar Perea, and Esteve Clua. 2007. *Corpus Oral Dialectal (COD)*. Barcelona: Departament de Filologia Catalana, Universitat de Barcelona.

Volpe, Mark Joseph. 2005. *Japanese morphology and its theoretical consequences: Derivational Morphology in Distributed Morphology*. Doctoral dissertation, Stony Brook University.

Watanabe, Akira. 2012. Person-number interaction: Impoverishment and natural classes. Ms., University of Tokyo.

Watkins, Laurel J. 1984. *A Grammar of Kiowa*. With the assistance of Parker McKenzie. Lincoln: University of Nebraska Press.

Wheeler, Max. 1998. Catalan. In Martin Harris and Nigel Vincent, eds., *The Romance Languages*, 170–208. London and New York: Routledge.

Wiese, Richard. 1996a. Phonological versus morphological rules: On German umlaut and ablaut. *Journal of Linguistics* 32:113–135.

Wiese, Richard. 1996b. *The Phonology of German*. Oxford: Oxford University Press.

Williams, Edwin S. 2006. Dumping lexicalism. In Gillian Ramchand and Charles Reiss, eds., *Oxford Handbook of Linguistic Interfaces*, 353–381. Oxford: Oxford University Press.

Wolf, Matthew. 2006. For an autosegmental theory of mutation. In L. Bateman, M. O'Keefe, E. Reilly, and A. Werle, eds., *University of Massachusetts Occasional Papers in Linguistics 32: Papers in Optimality Theory III*. Amherst: GLSA, University of Massachusetts.

Wood, Jim. 2012. Icelandic morphosyntax and argument structure. Doctoral dissertation, New York University.

Wunderlich, Dieter. 2012. Polarity and constraints on paradigmatic distinctness. In Jochen Trommer, ed., *The Morphology and Phonology of Exponence*, 160–194. Oxford: Oxford University Press.

Wurmbrand, Susi. 1999. Modal verbs must be raising verbs. In Sonya Bird, Andrew Carnie, Jason D. Haugen, and Peter Norquest, eds., *Proceedings of the 18th West Coast Conference on Formal Linguistics (WCCFL)*, 599–612. Somerville, MA: Cascadilla Press.

Wurmbrand, Susi. 2011. Tense and aspect in English infinitives. Ms., University of Connecticut. Storrs. http://wurmbrand.uconn.edu/Susi/Infinitives.html.

Wurmbrand, Susi. 2012. Agreement: Looking up or looking down? Lecture given in Agreement seminar, MIT, March 2012. http://wurmbrand.uconn.edu/Papers/MIT-2012.pdf.

Wurmbrand, Susi. Forthcoming. QR and selection: Covert evidence for phasehood. In Stefan Keine and Shayne Sloggett, eds., *Proceedings of the North Eastern Linguistics Society (NELS) Annual Meeting 42*. Amherst: GLSA, University of Massachusetts. http://wurmbrand.uconn.edu/Papers/NELS42.pdf.

Wurmbrand, Susi, Artemis Alexiadou, and Elena Anagnostopoulou. 2012. Disappearing phases: Evidence from raising constructions cross-linguistically. Paper presented at the 5th Conference on Syntax, Phonology and Language Analysis (SinFonIJA), Vienna, Austria, September 2012.

Yang, Charles. 2002. *Knowledge and Learning in Natural Language*. Oxford: Oxford University Press.

de Yrizar, Pedro. 1992. Morfología del Verbo Auxiliar Vizcaíno: Estudio Dialectológico. Bilbo: Euskaltzaindia. http://www.euskaltzaindia.net/mvav.

de Yrizar, Pedro. 2002. Morfología del Verbo Auxiliar Souletino: Estudio Dialectológico. Bilbo: Euskaltzaindia. http://www.euskaltzaindia.net/mvav.

Yu, Alan. 2004. The morphology and phonology of infixation. Doctoral dissertation, University of California, Berkeley.

Zubiri, Ilari, and Entzi Zubiri. 2012. *Euskal gramatika osoa*. Bilbo: Ikasbook.

Zwarts, Joost, Petra Hendriks, and Helen de Hoop. 2005. Comparative paths to an optimal interpretation. In Emar Maier, Corien Bary, and Janneke Huitink, eds., *Proceedings of Sinn und Bedeutung 9*, 553–563. Groningen: Groningen University.

Zwicky, Arnold M. 1977. Hierarchies of person. In W. A. Beach, S. E. Fox, and S. Philosoph, eds., *Papers from the Thirteenth Meeting, Chicago Linguistics Society* 13, 714–733. Chicago: University of Chicago.

Zwicky, Arnold M. 1985. Rules of allomorphy and syntax-phonology interactions. *Journal of Linguistics* 21:431–436.

Contributors

Karlos Arregi   University of Chicago

Jonathan David Bobaljik   University of Connecticut

Eulàlia Bonet   Centre de Lingüística Teòrica, Universitat Autònoma de Barcelona

David Embick   University of Pennsylvania

Daniel Harbour   Queen Mary University of London

Heidi Harley   University of Arizona

Alec Marantz   New York University/NYU Abu Dhabi

Tatjana Marvin   Univerza v Ljubljani

Ora Matushansky   CNRS/Université Paris VIII/UiL OTS/Universiteit Utrecht

Martha McGinnis   University of Victoria

Andrew Nevins   University College London

Rolf Noyer   University of Pennsylvania

Isabel Oltra-Massuet   Universitat Rovira i Virgili

Mercedes Tubino Blanco   Universidad de Sevilla

Susi Wurmbrand   University of Connecticut

Language Index

Marshallese, 138
Menominee, 136
Mokilese, 139
Mussau, 138

Nama, 56
Newari, 141
Nyawaygi, 139

Passamaquoddy, 57
Persian, 187–188, 196
Proto-Indo-European, 21–23, 25, 27, 29,
   31, 33, 35, 37

Romance, 2, 4, 17, 19, 143
Romanian, 141, 197
Russian, 21–23, 25–27, 29, 35, 65, 68,
   77, 196

Sanskrit, Vedic, 21–22, 35, 37
Semitic languages, 54–55
Southern Tiwa, 150
Spanish, 4, 13, 18–19, 132, 145, 155,
   164, 167, 171, 173–176, 178–179,
   181–183, 197, 207, 212
Surmiran, 19
Sursurunga, 137

Terena, 152, 154
Tewa, 136, 143–148, 150
Tibetan, 141
Tibeto-Burman, 141
Tiwa, Southern, 150
Tok Pisin, 138
Tsafiki, 141

Ubykh, 187, 196

Walapai, 139

Yaqui. *See* Hiaki
Yimas, 137–138, 143
Yoeme. *See* Hiaki
Yucatec Mayan, 54

Zulu, 164